Archeofuturism

ARCHEOFUTURISM
EUROPEAN VISIONS OF THE POST-CATASTROPHIC AGE

Guillaume Faye

ARKTOS

First English edition published in 2010 by Arktos Media Ltd.
Copyright to the English edition © 2010 by Arktos Media Ltd.

All rights reserved. No part of this book may be reproduced or utilised in any form or by any means (whether electronic or mechanical), including photocopying, recording or by any information storage and retrieval system, without permission in writing from the publisher.

Printed in the United Kingdom

ISBN **978-1-907166-09-9** (Softcover)
ISBN **978-1-907166-10-5** (Hardcover)

BIC classification: Social & political philosophy (HPS);
General & world history (HBG)

Translation: Sergio Knipe
Editor: John B. Morgan
Proofreader: Michael J. Brooks
Cover Design: Andreas Nilsson
Layout: Daniel Friberg

ARKTOS MEDIA LTD

www.arktos.com

Table of Contents

Foreword by Michael O'Meara 7

A Note from the Editor 11

Introduction 13

1. An Assessment of the Nouvelle Droite 23

2. A Subversive Idea: Archeofuturism as an Answer to the Catastrophe of Modernity and an Alternative to Traditionalism 53

3. Ideologically Dissident Statements 91

4. For a Two-Tier World Economy 161

5. The Ethnic Question and the European 175

6. A Day in the Life of Dimitri Leonidovich Oblomov –
   A Chronicle of Archeofuturist Times 195

Foreword

*'We have kept faith with the past,
and handed down a tradition to the future.'*
– Patrick Pearse, 1916

Guillaume Faye was long associated with that school of thought, which, in 1978, the French media labelled 'la Nouvelle Droite' – though it was Right wing in no conventional sense, representing, as it did, the distinctly postmodern cause of 'European identitarian nationalism'.

Not to be confused, then, with the various neo-liberal, implicitly Protestant, and market-oriented tendencies bearing the same designation in the English-speaking world, the French New Right grew out of GRECE (the Groupement de Recherche et d'Études pour la Civilisation Européenne), an association formed in 1968 by various anti-liberals hoping to overcome the failed legacies of Pétainism, neo-fascism, Catholic traditionalism, regionalism, colonialism, and Poujadism – in order to resist the cancerous Americanisation of their homeland.

To this end, GRECE's founders believed they would never overthrow America's liberalising hegemony, as long as the general culture remained steeped in liberal beliefs. In the formulation of its master thinker, Alain de Benoist: 'Without Marx, no Lenin'.

That is, without the ascendence of anti-liberal ideas in the general culture and thus without a revolution of the spirit, there would be no viable movement against *le parti américain*.

GRECE was established, thus, not for *la politique politicienne*, but for the sake of metapolitically rearming European culture.

And in this, it was not unsuccessful. For GRECE's philosophically persuasive revival of anti-liberal thought and the subsequent affiliation of several prominent European thinkers to its banner made it an influence of some immediate import. Indeed, it can almost be said that for the first time since the Action Française, 'Rightists' in the '70s achieved a level of sophistication and attraction nearly 'comparable' to that of the Left, as France's 'intellectual right' threw off the defenceless conservatism that came with Americanisation to challenge the liberal consensus imposed in 1945.

* * *

While still working on his doctorate in Political Science at the elite Institut d'études politiques de Paris (Science Po), Guillaume Faye began gravitating to GRECE. By 1973, he had become its 'number two' advocate, a role he would play until 1986.

Like other *Grécistes* in this early period, Faye was influenced by those European currents that had previously countered the imposition of liberal ideology.

Foremost of these counter-currents were the Conservative Revolution of the German 1920s (Spengler, Moeller van den Bruck, Schmitt, Freyer, Heidegger, Jünger, etc.); the traditionalism of Julius Evola; the Indo-Europeanism of Georges Dumézil; and the heritage of pre-Christian paganism.

Contemporary anti-liberal ideas in stream with these deeper currents — such as the ethology of Konrad Lorenz, the philosophical anthropology of Arnold Gehlen, or the illusion-destroying field of genetics — were similarly incorporated into GRECE's anti-liberal curriculum.

Faye, though, took to these ideas differently (more radically, in my view) than de Benoist — perhaps because of his earlier affiliation with the Situationists and the 'aristocratic' ex-Communist Henri Lefebvre; more probably because of his apprenticeship with the Italian journalist, Germanist, and post-fascist firebrand Giorgio Locchi; and ultimately, of course, because of his specific temperament.

Less prolific and encyclopedic than de Benoist, the younger Faye was considered by some the more creative (*le véritable moteur intellectuel de la nouvelle droite*). He played second fiddle, though, to the master, who seemed bent on blunting the edge of New Right radicalism. There was, as a consequence, a certain implicit tension between their different notions of the anti-liberal project.

FOREWORD

* * *

For reasons explained in the first chapter, Faye quit GRECE in 1986. During the next dozen years, he worked in the 'media' as a radio personality, journalist, and occasional ghost writer.

The publication of *L'Archéofuturisme* in 1998 signaled his return to the metapolitical fray.

At one level, this work accounts for the dead-end that de Benoist's GRECE had got itself into by the mid-1980s, suggesting what it could have done differently and with greater effect.

At another, more important level, it addresses the approaching *interregnum*, endeavouring to 'transcend' the historical impasse, which pits the ever changing present against the heritage of the past, between European modernism and traditionalism.

To this end, Archeofuturism calls for 'the re-emergence of archaic configurations' – 'pre-modern, inegalitarian, and non-humanist' – in a futuristic or long-term 'context' that turns modernity's forward, innovative thrust (totally nihilistic today) into a reborn assertion of European being, as the temporal and the untimely meet and merge in a higher dialectic.

Archeofuturism is thus both archaic and futuristic, for it validates the primordiality of Homer's epic values in the same breath that it advances the most daring contemporary science.

Because the Anglophone world outside the British Isles is a product of liberal modernity, the struggle between tradition and modernity, pivotal to continental European culture, has been seemingly tangential to it.

This struggle, however, nevertheless now impinges on the great crises descending on the U.S. and the former White dominions.

Faye's Archeofuturism holds out an understanding of this world collapsing about us, imbuing European peoples with a strategy to think through the coming storms and get to the other side – to that post-catastrophic age, where a new cycle of being awaits them, as they return to the spirit that lies not in the past per se, but in advance of what is to come.

– MICHAEL O'MEARA
Saint Ignatius of Loyola Day, 2010

Michael O'Meara is the author of two vital books on the subject of the New Right in English, *New Culture, New Right* and *Toward the White Republic* and has published many articles on related topics.

A Note from the Editor

As Faye's text did not originally contain any footnotes, all of those present in this edition were added by myself. Faye was writing for a primarily French audience who could be expected to be familiar with the many figures and concepts from the French New Right and from French literary, political and philosophical history to which he refers, but which may be unfamiliar to an English-speaking reader. As such, I have added footnotes where I felt they could serve to explicate such references. Likewise, Faye occasionally makes reference to contemporary political events of which the average reader might have been aware in 1998, but which may be unfamiliar today, more than a decade later. I have therefore added details about some of these events where I felt it was appropriate.

– JOHN B. MORGAN

Introduction

The thread of this book is formed by three logically connected theses. The first argues that current civilisation, a product of modernity and egalitarianism, has reached its final peak and is threatened by the short-term prospect of a global cataclysm resulting from a *convergence of catastrophes*. Many civilisations have disappeared in the past, but these were disasters that only affected certain areas of the Earth, not the whole of humanity. Today, for the first time in history, a world civilisation – the global extension of Western civilisation – is threatened by converging lines of catastrophe produced by the implementation of its ideological plans. A dramatic chain reaction of events is converging towards a fatal point which I believe may occur in the early Twenty-first century, between 2010 and 2020. This will plunge the world as we know it into chaos and cause a genuine cultural earthquake. These 'catastrophe lines' concern the environment, demography, economy, religion, epidemics and geopolitics.

The present civilisation cannot endure. Its foundations are contrary to reality. It is clashing not so much against ideological contradictions – which can always be overcome – but, for the first time, against a physical wall. The old *faith in miracles* of egalitarianism and the philosophy of progress – which suggests one can have his cake and eat it too – is now coming to an end. This fairytale ideology has led to a world of illusions that is less and less credible.

Second thesis: the individualist and egalitarian ideology of the modern world is no longer suitable in an increasing number of spheres in our civilisation. To face the future, it will be more and more necessary to adopt an archaic mind-set, which is to say a pre-modern, non-egalitarian and non-humanistic outlook – one capable of restoring the ancestral values that inform 'orderly societies'. Already the advancements made in technology and science, particularly in biology and computer science, can no longer be managed with modern humanistic values and ways of thinking; already geopolitical and social events point to the tumultuous and violent emergence of problems connected to religion, ethics, food production and epidemics. It is necessary to return to primary issues. Hence the new idea I am suggesting: *Archeofuturism*. This idea enables us to make a break with the obsolete philosophy of progress and the egalitarian, humanitarian and individualist dogmas of modernity, which are unsuited to our need to think about the future and survive the century of iron and fire that is looming near.

Third central thesis: we should already envisage the aftermath of the chaos, the *post-catastrophic world*, according to the principles of Archeofuturism, which are radically different from those of egalitarian modernity. This book gives an outline of them. It is useless to try and conceive reforms inspired by provisional wisdom and rationality: man is incapable of doing so. Only when man finds himself with *his back against the wall*, in an emergency, does he react. What I will offer here is a sort of mental training for the post-catastrophic world.

* * *

The term 'Conservative Revolution', which is often used to describe our current of thought, is not enough. The word 'conservative' has demobilising, anti-dynamic and rather outdated connotations. Today it is not a matter of 'conserving' the present or returning to a recent past that has failed, but rather of regaining possession of our most archaic roots, which is to say those most suited to the victorious life. One example, among others, of this inclusive logic: to synthesize technological science and archaism – to reconcile Evola[1] and Marinetti,[2] Doctor Faust and the Labourers.[3]

---

1   Julius Evola (1898-1974) was the most important Italian member of the traditionalist school, which is to say that he opposed modernity in favour of an approach to life consistent with the teachings of the ancient sacred texts. Toward this end he also became intellectually involved with the Fascist and National Socialist movements in their heyday (although considered both to be woefully deficient). His most important book, available in English, is *Revolt Against the Modern World*.

2   F. T. Marinetti (1876-1944) was the founder of the Futurist movement in Italian art, which derided tradition in favour of technology and social change.

3   Faye is likely referring to a famous section of a longer poem, *Jocelyn*, by this name which was written in 1801 by the French poet Alphonse de Lamartine (1790-1869). De Lamartine is

INTRODUCTION

The controversy between 'traditionalists' and 'modernists' has grown barren. We should be neither of these things, but rather Archeofuturists. Traditions are made to be purged, screened and selected: for many of them breed viruses, of the kind that are exploding today. As for modernity, it probably no longer has a future.

The world of the future will be precisely as Nietzsche and the great but unjustly – or perhaps justly – ignored philosopher Raymond Ruyer[4] foresaw it.

In this book, I also aim to positively define the flexible and rather neutral concepts of 'post-modernity' and 'anti-egalitarianism' by constructing a new term that describes an ideology to be developed: *vitalist constructivism*.

'Convergence of catastrophes', 'Archeofuturism', 'vitalist constructivism': I have always tried to come up with new concepts because only through ideological innovation can we avoid rigid and obsolete doctrines in a world that is rapidly changing and where dangers are taking shape. In such a way, an idea armed with ever new weapons can win the 'war of words', shock reality and stir people's conscience.

I am showing some *paths*, not formulating dogmas; my aim is not so much to assert my theses (which belong to what Socrates called *doxa* – 'opinion' that is open to question), but rather to launch a debate on crucial problems, in such a way as to make a break through the ideological insignificance, blindness and poverty that have intentionally been created by the system to divert people's attention and conceal its own complete failure. In a society that considers all genuine *ideas* subversive, which seeks to discourage ideological imagination, and which aims to abolish thought in favour of spectacle, the main goal must be to awaken people's consciences, raising traumatising problems and sending ideological electroshocks: *shocking ideas*.

* * *

---

regarded as the first poet of the Romantic school in France. 'The Labourers' describes the life of common farmers, depicting their family and working lives as being in accordance with the glory of God and the natural world. It has been translated by F. H. Jobert in *Jocelyn: An Episode* (London: Edward Moxon, 1837), pp. 358-375. It is worth noting that de Lamartine was a Pantheist and regarded Islam as the greatest religion.

4 Raymond Ruyer (1902-1987) was a French philosopher who wrote primarily about the philosophical implications of the discoveries of modern science and his own form of gnosticism. He opposed existentialism and the Leftist trends in the philosophy of his time. He has never been translated into English and, as Faye writes, is largely forgotten in France today. Faye will discuss Ruyer at greater length later in this book.

I did not want to write a traditional essay, divided into chapters and with a rather cumbersome structure, so I proceeded through glimpses and sketches, each shedding more or less light, to make the book easier to read. Besides, I do not strictly confine my discussion to its central theme, but also seek to engage with related issues such as the crucial problem of the current demographic colonisation of Europe by Afro-Asiatic peoples, and which is prudishly called 'immigration'. Towards the end of the book the reader will find a futuristic political fiction that will immerse him in the Archeofuturist post-catastrophic world, in the year 2073, at the heart of the Eurosiberian Federation.

* * *

We should make a break with soft ideas, now that the real issues are becoming central again. Some people may regard many of my suggestions as *ideologically delinquent* in the context of the ruling ideology and pseudo-virginal chorus of the self-righteous. Well, they are.

You may wonder why I have not published any ideological texts in thirteen years and have only now resumed my battle of ideas. It is mainly because after spending a long time 'with the enemy', I have understood many things and have been able to renew and adjust my position. To be radically opposed to a given model of society, it is necessary to know it well, from the inside. It is always very interesting to stand at the heart of the military apparatus of the enemy, to be *in* the world without being *of* the world: the cobra tactic.

Moreover, the growing stakes and increasing gravity of the signs that herald imminent catastrophes have led me to return to the battlefield and revise many positions I had once adopted, when active in the Nouvelle Droite,[5] in order to seek paths more appropriate for the 'exceptional case' (Carl Schmitt's *Ernstfall*[6]) we are currently facing. No

---

5 New Right.
6 Carl Schmitt (1888-1985) was an important German jurist and legal philosopher who was part of the Conservative Revolution of the Weimar era. *Ernstfall*, one of his key concepts, is often translated as 'state of exception' or 'emergency case'. Schmitt's use of this concept is complex, but in brief, Schmitt regarded the rule of law in any society as always being a temporary state of affairs and that modern, liberal concepts of law in particular are insufficient when confronted with a situation that falls outside the routine situations which they were designed to regulate. As such, it is the responsibility of the leaders of a society to determine when the law must be suspended in order to deal with an exceptional situation. Schmitt regarded the National Socialists' abrogation of the Weimar constitution as being a legitimate use of the *Ernstfall*. Schmitt discusses this idea at length in his book *Political Theology*.

doubt, the new courses I am suggesting we should take are far more radical than those I promoted thirteen years ago – 'radical' being a synonym not of 'extremist', but of 'fundamental'.

Our current of thought is being offered a real historical chance, for: first, facts are proving us right; second, the global system established by our ideological enemy is about to collide with the wall of reality and plunge into the abyss, both in France and worldwide; and third, the ruling ideology has nothing new to offer – no solutions – unless it contradicts itself. Its only answer consists of simulacra and pretences, in an attempt to make people forget and to divert their attention: what Guy Debord[7] described as the strategy of 'spectacle' at a time when this was still going strong. Instead, today, despite being a thousand times more sophisticated, this strategy is seizing up and shaking like a motor with an empty tank. We are facing a deafening ideological silence, made of worn-out, softened values and a lack of conviction in their own beliefs. Nor do establishment intellectuals have any intellectual Viagra to get some stimulation. This is a critical combination of circumstances which we should grab by the hair.

* * *

We must take up the idea of *Revolution* again, a notion that has been misinterpreted and betrayed by the charlatans of the Left for over two centuries. Once, the newspaper *Combat*[8] used the nice slogan 'from Resistance to Revolution'. It is not a matter of simply resisting the destruction that unfolds before our eyes and is spreading with a power we find hard to conceive, but rather of envisaging an 'aftermath of the system', on the basis of a worldview (and of the ideologies and doctrines stemming from it and which it will be worth describing) that is genuinely revolutionary: a worldview, that is, which makes a radical break with contemporary values and morals, in order to

---

7   Guy Debord (1931-1994) was a French Marxist philosopher and the founder of the Situationist International whose ideas have become influential on both the radical Left and Right. The spectacle, as described in his principal work, *The Society of the Spectacle*, is one of the means by which the capitalist establishment maintains its authority in the modern world – namely, by reducing all genuine human experiences to representational images in the mass media, thus allowing the powers-that-be to determine how individuals experience reality.

8   *Combat* was originally an underground newspaper published by the French Resistance during the German occupation of the Second World War. Many Left-wing intellectuals who would become highly influential in the post-war period wrote for it, including Jean-Paul Sartre, Albert Camus, André Malraux and Raymond Aron. The paper continued to function for many years after the war as a mouthpiece for the French Left.

train spirits for the world of the future and create active minorities ready to experience this break and adopt an Archeofuturist ethic with detachment.

Our current of thought, broadly conceived, must necessarily unite on a European level, forgetting about provincial disputes and narrow doctrines, in such a way as to seize the opportunity it is being offered: to acquire the *monopoly over alternative thought* – rebel thought. Let us take advantage of the present global crisis and formulate suggestions that may stir the conscience of the young.

We should avoid being backward-looking, concerned with restoration and reaction, for it is the last few centuries that have spawned the pox that is now devouring us. It is a matter of returning to archaic and ancestral values, while at the same time envisioning the future as something more than a mere extension of the present. Against modernism, futurism. Against attachment to the past, archaism. Modernity has failed, it is crumbling, and its followers are the real reactionaries.

* * *

We are standing *face to face with the barbarians*. The enemy is no longer outside but inside the City, and the ruling ideology, paralysed, is incapable of spotting him. It stammers, overcome by its own moral disarmament, and is giving up: this is the time to seize the reins. Present society is an accomplice to the evil that is devouring it. For this reason, if the ideas our current of thought espouses will prove an effective alternative, they will be accused by the chorus of false virgins of two demonising anathemas: subversion and sedition. Why not?

We should expect this. We should engage in battle without complaining about censorship and persecution, and without being surprised, should the ruling ideology betray its own principles to fight against its absolute enemy.

With respect to the system, and especially the intellectual Left – its most faithful guard dog – our current of thought and its associated political forces now find themselves in much the same situation Leftists and anarchists were facing in May '68[9] with respect to the

---

9 May 1968 was when a series of strikes by radical Left-wing student groups in Paris, under the influence of Guy Debord and the Situationist International, were joined by a strike by the majority of the French work-force, shutting down France and nearly bringing down the government of Charles de Gaulle. Although the strikes ended in failure and had evaporated by July, they are still seen as the decisive moment when traditional French society, including the old Leftist and

INTRODUCTION

establishment. Still, there are some considerable differences: on the one hand, radical Leftists and anarchists at the time were leading a struggle for workers' empowerment, a backward-looking, symbolic battle with no real stakes; on the other, more traditional Leftists and the Right-wing ruling power at the time ultimately shared the same egalitarian ideology, while disagreeing as to how and to what extent this should be applied. As for the far Left of today, as we shall see, it serves to accelerate official ideology and praxis, while concealing the role it plays through pseudo-dissent: actually, in no way does it challenge the dominant global model of civilisation or economy.

* * *

By contrast, the situation our forces find themselves in with respect to the system is similar to that which existed in the 1930s: no point of agreement is possible (except on the part of the potential traitors of the parliamentary Right, which form a rather significant portion of the ruling class): the only strategy is all-out war. In adopting a revolutionary stance, aimed at the overthrowing of a civilisation, we must be ready to face total war – a fight without quarter. Clearly, the enemy will seek to get rid of us by any means, just as we will have to make sure that his return to the political scene is made utterly impossible.

As Hölderlin's famous verse goes, 'This is the midnight hour of the world.'[10] And when the sun rises, the morning will have to belong to us. Giorgio Locchi[11] used to say much the same thing: we are living in the *interregnum* between the collapse of a system and the creation of the new metamorphic universe.

There is a present need to develop a worldview that may serve as the common denominator for our current of thought on a European level, and which in the face of an emergency may enable us to overcome minor disputes caused by differences in doctrine or attitude. The

---

Communist parties, were forced to give way to the more liberal attitude that has come to define France in subsequent years.

10 The author is here most likely referring to Hölderlin's poem 'Bread and Wine'. The night is used to symbolically represent our age, when the ancient gods of Greece and Christ have left the world and it is only the poets who attempt to keep their memory alive until their return. Many translations exist. Martin Heidegger discusses this poem at length in his famous essay 'Why Poets?', translated in *Off the Beaten Path* (Cambridge: Cambridge University Press, 2002).

11 Giorgio Locchi (1923-1992) was an Italian journalist who was a founding member of GRECE and an occasional collaborator with Alain de Benoist. He also wrote on Wagner and Nietzsche. He remains untranslated.

notion of Archeofuturism may certainly contribute to this. As Nietzsche already prophesised, 'The man of the future is he who will have the longest memory.'[12]

* * *

Clearly, I remain loyal to the overall notion of 'nationalism', understood however in its European, continental understanding as opposed to the French, which we have inherited from the dubious philosophy of the Revolution. To be a nationalist today is to assign this concept its original etymological meaning, 'to defend the native members of a people.' This entails a break with the traditional idea of nation and citizenship we have inherited from the egalitarian philosophy of the Enlightenment. To be a nationalist today is to embrace the notion of a 'European people', which exists and is under threat, but is not yet politically organised for its self-defence. It is possible to be a 'patriot', someone tied to his sub-continental motherland, without forgetting that this is an organic and vital part of the common folk whose natural and historical territory – whose fortress, I would say – extends from Brest to the Bering Strait.

It is quite true that the form of present-day Europe, this 'thing', must be fought. Yet, the *historical* tendency of the European peoples to unite in the face of adversity must be defended to the very end. Some of my positions in this book, in favour of the establishment of a United States of Europe or Eurosiberian Federation, may shock certain people. But let there be no doubt: I am not a partisan of the spineless Europe of the Amsterdam Treaty, nor am I an enemy of France. Again, what I am suggesting here are paths: I am providing weapons to launch the debate and trying to point to some 'value guidelines' – in no case am I offering a closed doctrine.

The European youth – the genuine one – needs ideas to face the imminent danger, not video-centric revelries or humanitarian whimpers in a climate of sophisticated censorship and repression. The 'Mitterrand generation' is dead, engulfed by ridicule and paralysed by failure. Now is the time for a *dissident generation* to rise. It is up to her to imagine the unimaginable.

---

12 This quote is the motto of Terre et Peuple, a group composed of intellectuals who have broken away from GRECE or the Front National. Faye has contributed to their journal.

INTRODUCTION

* * *

If it is to survive, our folk – whether in Toulouse, Rennes, Milan, Prague, Munich, Antwerp or Moscow – must revert to and embrace ancestral virility. Otherwise, as is already happening, we shall be submerged by more vital, more youthful and less well-meaning peoples with the complicity of a delinquent bourgeoisie that – whatever it may do – will itself be swept away by the tide it has so heedlessly caused.

Let us dare to think the unthinkable. Let us explore and continue along the path paved by an early riser and visionary: a certain Friedrich Nietzsche. From Resistance to Revolution, from Revolution to Rebirth.

1

AN ASSESSMENT OF THE NOUVELLE DROITE

Why did I suddenly quit the Nouvelle Droite and its most precious flower, GRECE,[1] in 1986? The answer is a very simple one. No, I was not recruited by the CIA, nor did I lose my mind through the bite of a rock'n'roll-singing mosquito. Firstly, some work projects prevented me from contributing to the activities of GRECE as a militant; secondly, I noticed that the tone and general orientation of the movement were losing momentum and turning it into a kind of clique and club. And thirdly, the Nouvelle Droite was taking ideological turns with which I increasingly disagreed and which threatened to marginalise it, despite the (always verifiable) worth of its members – and there was nothing I could do to change its course. Twelve years later my diagnosis has been confirmed: the influence of the Nouvelle Droite has declined. Why?

Diagnosis: A Considerable Loss of Influence

Once, every issue of *Éléments*[2] was an ideological barrage that sparked outraged reviews from the mainstream press. Today, the magazine has taken on an almost secretive tone and is ignored by the wider educated public and the decision-makers. Likewise, the '*Colloques parisiens*'[3] no longer receive the media coverage they enjoyed in the 1980s. While

---

1 Groupement de recherche et d'études pour la civilisation européenne, or 'Group for the Research and Study of European Civilisation'. Founded in 1968 by Alain de Benoist, it has always been the most prominent group associated with the Nouvelle Droite, or French New Right.
2 *Éléments*, along with *Krisis* and *Nouvelle École*, are the official journals of GRECE.
3 Paris conferences.

they may still attract roughly the same number of people, have they not become the nostalgic meetings of an association of veterans? Besides, I doubt whether GRECE is still as capable as it was in previous years of filling the halls of large cities in France and Belgium with weekly conferences and seminars. The only recent instance in which the Nouvelle Droite had any relevance was the debate launched on *Krisis* magazine regarding the fraud of contemporary art: a central problem that shocked the little subsidised masters and gigolo-artists of mainstream *non-art*. Alas, this media visibility was short-lived and insufficient: ultimately it was largely ignored by the wider public, unlike the heated polemics on *central issues* we continued to spark up until the mid-1980s, and which spread everywhere, from the United States to the USSR.

Today even the most interesting writings from the Nouvelle Droite only circulate in the narrow milieu of its followers, while the platitudes, virtuous and verbose inanities, and self-righteous quibbles of people like Ferry,[4] Serres[5] and Conte-Sponville,[6] just like Bourdieu's[7] idiocies and the talentless gloom of Bernard-Henri Lévy[8] – mediocre mediatised intellectuals sponsored by the current soft totalitarianism – are spreading through the insolent self-importance of idiots. This is a defeat. But losing a battle does not necessarily mean losing the war.

In brief, the Nouvelle Droite has been confined to the *periphery of the debate*. Regrettably, it has turned into an ideological ghetto. It no longer sees itself as a powerhouse for the diffusion of energies with the ultimate aim of acquiring *power*, but rather as a publishing enterprise

---

4   Luc Ferry (1951-), a French philosopher who advocates secular humanism. From 2002-04 he was the Minister of Education.

5   Michel Serres (1930-) is a very prominent French philosopher who frequently writes about the philosophy of science.

6   André Comte-Sponville (1952-) is a French philosopher who advocates a spiritual form of atheism.

7   Pierre Bourdieu (1930-2002) was a prominent French anthropologist, philosopher and sociologist who studied social dynamics, and he opposed neo-liberalism and globalisation.

8   Bernard-Henri Lévy (1948-) is a French philosopher of Jewish ethnicity who was initially known for his rejection of Marxist beliefs which had become commonplace in France by the 1970s. In more recent years he has become best known for his opposition to Muslim influence on European culture and his support for the Iraq War, and for his book *Who Killed Daniel Pearl?*, in which he claims that Pearl was killed because he had learned too much about the connections between Al Qaeda and the Pakistani government. Although popular, Lévy has frequently been criticized by other French intellectuals for his methods and egotistical style of presentation. In 2010, he was publicly humiliated when it was revealed that an essay he had written in an effort to refute Kant had based its arguments upon the ideas of a philosopher who was a fictional character created as a parody by a French journalist.

AN ASSESSMENT OF THE NOUVELLE DROITE

that also organises conference but has limited ambitions. Clearly, behind this process of marginalisation lie both external causes (a hostile or indifferent milieu) and internal ones (due to the movement itself). The latter are more crucial. One can only recover from a temporary defeat by acknowledging it as such and assuming responsibility for it. Ambition comes from modesty: no progress can be made without self-criticism. Those who blame others, enemies and the political climate for their own failures do not deserve to win. For it is in the logic of things for enemies to oppress you and circumstances to prove hostile. The mistake lies in exorcising reality by adopting the morals of intention as opposed to those of consequences, through unrealistic arguments: 'You know, we have as many people as before at the *colloques*'; 'It is full of young people at the Université d'été.'[9] Hell! We should stop issuing reassuring reports that only serve to conceal reality. It is necessary to avoid sterile polemics and accept *positive self-criticism*. The question is: why is the Nouvelle Droite, which possessed an impressive ideological armoury, objectively going downhill? Are we witnessing its final decline or merely a standstill that foreshadows its relaunching?

I will attempt to answer this question, but first, two preliminary observations are in order. The first is that no one, within what might vaguely be defined as the 'ideological Right' in Europe, has yet managed to acquire the kind of intellectual influence the Nouvelle Droite had at the turn of the 1980s. Its only potential heir is the pan-European intellectual movement 'Synergies',[10] which is led – amongst others – by Robert Steuckers, and which strikes me as being on the right track, for it pursues ambitious aims. Still, the match is not over. A second observation: in 1998, the *only* genuinely tangible influence of the Nouvelle Droite on society at large is that exercised by its runaway members now in the Front National,[11] which they have driven to take an anti-American course – a real mental revolution for this milieu. On the other hand, the influence of the Nouvelle Droite can be detected in the formulation of a widespread cultural and economic hostility towards Americanisation ('the French exception') – hostility which remains

---

9 Summer University. This is a course that the Nouvelle Droite used to run.
10 This group still exists. They maintain a Web site at http://euro-synergies.hautetfort.com/.
11 The Front National is a far-Right nationalist party which was founded by Jean-Marie Le Pen in 1972. Over the last decade, the FN has had a number of significant electoral successes. Le Pen remains its leader.

largely ineffectual, given the indolence of political decision-makers. So overall, the concrete ideological impact of the Nouvelle Droite has been rather meagre.

From 1986 I started to feel that there was no longer any real fervour left, and that a clique spirit and literary pagan romanticism were prevailing over historical will. I could see that the chief aim consisted no longer in the establishment of a school of thought, the exercising of concrete ideological influence, and the development of a radical thought through 'shocking ideas', but rather in a sort of elegant intellectualism and the entrenchment of a 'community', a noble thing when it relies on an established power, but a demobilising one when reduced to the tautology of a clique.

It is necessary to analyse the causes of this decline, which – having taken place in less than a decade – was far more sudden and striking than the old Action Française...[12] How and why did the main alternative ideological movement to have emerged in post-War Europe turn out to be merely a comet? What lessons can we gain from this? And what should we do now? Can the mechanism be put into motion again?

Certainly, no one knows what will remain in future history of the mass of texts produced by the Nouvelle Droite and its followers. No doubt, there will be some continuations, restatements, and reinterpretations. Perhaps a revolution in 2050? But let us simply stick to the present for the time being, before moving on to discuss the strategies for a restoration.

The Causes of the Loss of Influence of the Nouvelle Droite

It is quite true that cultural societies, theoretical magazines, and new intellectual systems must all face great obstacles which did not exist only twenty years ago: the end of the pyramidal spread of knowledge, the firepower of the cultural entertainment industries that marginalise and conceal all new or rebellious thought, the net-like multiplication

---

12 The Action Française was a Right-wing monarchist group founded in 1898 which enjoyed a great deal of support. During the 1920s and '30s, it became increasingly favourable towards Fascist regimes, and the French government, feeling threatened, banned it along with other Far Right groups in 1936. The AF experienced a revival under the Vichy regime during the German occupation, when it supported the Vichy government, causing its dissolution yet again after the liberation of Paris in 1944.

AN ASSESSMENT OF THE NOUVELLE DROITE

of medias of all sorts, etc. These external causes, however, do not explain everything. The Nouvelle Droite might have turned obstacles into opportunities by adapting its communication strategy to the new milieu. It failed to do so – *we* failed do so. We failed to see the meteor that was approaching.

I believe that the chief causes for this withdrawal are:

1 – The competitive emergence of the Front National and of the thought of Antonio Gramsci,[13] which was badly understood by the Nouvelle Droite.

2 – A tightening of censorship through a blacking out and closing off of the mainstream media, which followed the strengthening of ideological interdictions against all forms of alternative thought: the Nouvelle Droite submitted to these diktats, not daring to fight them through a creative, *disorienting* and provocative reaction.

3 – The profound inadequacy of Nouvelle Droite publications with respect to current media communication strategies, combined with an editorial tactic that was hardly effective.

4 – The continued adoption of an outdated 'apparatus logic' of the type to be found in political parties, which was not appropriate for a movement and school of thought, as well as journalistic or editorial policy, and which led cadres to flee on account of 'problems with the apparatus'.

5 – A certain ideological fossilisation, combined with the persistence of a 'Rightist cultural attachment to, and sentimentalising of the past' and the abandonment, in many fields, of the idea of 'radical thought' – the only kind of thought capable of sending a shockwave to arrest the media black-out. To this we can add the contradiction between implicit Euro-imperial references and an explicitly 'ethnopluralist' or even immigrationist stance.

6 – A (previously unknown) doctrinal softening on economic and scientific matters and a burgeoning of literary discourse.

7 – The favouring of criticism over positive formulations, of reaction over action.

Let us examine some of these points.

---

13 Antonio Gramsci (1891-1937) was an Italian Communist who was imprisoned by the Fascists. He developed the theory of cultural hegemony, which (in brief) holds that a political group cannot maintain power without first persuading the members of a society that the ideas it propagates are the normal state of affairs, thus giving itself legitimacy. Therefore, control over the cultural apparatus of a society is a prerequisite for holding power, rather than being something which follows a revolution. This idea has been highly influential on the European New Right.

1. The Front National and the 'Gramscian' Strategy

At first sight, the Front National could not have been a rival of the Nouvelle Droite, which never presented itself as being French nationalist. Yet different 'airlocks' exist in the family of the Right. The more ideologically unrefined public (or clientele?) is always attracted by the strongest pole. In the early 1980s, GRECE was the most important organisation in this area of politics: the Front National was considered a micro-group of good-for-nothings. We used to see them as being bigoted, papist, reactionary, servile towards America, jingoist and anti-European. Le Pen – this pirate-faced, confusion-stirring, neo-boulangist[14] old soldier – was barred from our meetings.

Then, by a twist of history, everything changed: the Front gained irresistible ascendency, and GRECE was no longer the pole of attraction that monopolized the movement. Like water leaking from a tap, cadres and leaders, even at the cost of ideological revisionism (something all too human), moved to *where something was happening*: the Front National. Bardet, Blot, Le Gallou, Martinez, Mégret, Millau, Vial and twenty others or so – all worthy men who were closely connected to GRECE or otherwise involved with it – transferred their skills to the Front National. Had it never appeared, it is likely that important 'human resources' would have remained in the sphere of the Nouvelle Droite. A veritable flight of brains…

Another reason why the Front caused the decline of GRECE is the former's *enticing of the media*, a phenomenon admen know well. The media, fascinated by the shocking political incorrectness of the Front National and of its President, soon forgot all about the Nouvelle Droite, which produced texts and events that were less attractive and provoking. Since the late 1980s, the Front has served as a *media screen* for the Nouvelle Droite, which – as we shall see – has proven incapable of reacting and opening counter-fire.

It should also be said that one of the handicaps of the Nouvelle Droite has been a poor reading of Gramscism, based on the adoption of the 'all is cultural, all is intellectual' strategy.

---

14 This refers to French General Ernest Boulanger (1837-1891), who served as War Minister and gained a popular following due to his advocacy of revenge against Germany for France's defeat in the Franco-Prussian War of 1870-71, conservative constitutional reforms and a restoration of the monarchy. After nearly provoking a war with Germany in 1887, Boulanger was dismissed. He and his followers threatened a coup in 1889, but Boulanger missed the opportunity due to his wish to take power legally, giving his opponents the time they needed to build a case against him and issue a warrant for his arrest. Boulanger fled the country and eventually committed suicide.

AN ASSESSMENT OF THE NOUVELLE DROITE

In our *metapolitical* 'Gramscian' strategy, we had simply overlooked the fact that the cultural battle Gramsci promoted was associated with the political and economic battle of the Italian Communist party, and as such did not take place 'in the void'. But unfortunately we had never actually read Gramsci... Ours was only braggadocio, pseudo-Gramscism. In order to prove effective, ideological and cultural action must be supported by concrete *political* forces which it integrates and extends. Chevènement's former CERES,[15] for instance, a satellite of the PS, or SOS-Racisme, another of its satellites, are clear examples of successful propaganda. In defining the founding idea behind the Nouvelle Droite in the 1970s, we had simply *underestimated the political element*.

By overestimating the cultural and intellectual pole, through a distorted analysis (of the works of Augustin Cochin[16]) and which found inspiration in cultural circles from before the French Revolution, we all too soon buried what would have been – and still is – a winning political strategy, without grasping the contemporary formula 'intellectual and cultural propaganda combined with electoral and political mobilisation'. We had forgotten we were no longer living in the Eighteenth century: that elections of various sorts were taking place every six months, and that politicians were the media heralds of a party system. The 'all is cultural' strategy only worked for the non-elective regimes of the past... We had announced the death of politics all too soon. Evidence for this is the fact that *Libération*[17] is more concerned by the mediatisation of Pierre Vial's association Terre et Peuple, a cultural and intellectual movement coordinated with the activities of a party, the Front National, than it is by Madelin[18] and Juppé's[19] circles of buddies.

---

15 Jean-Pierre Chevènement (1939-) was the founder of the Centre d'études, de recherche et d'éducation socialistes, or CERES (Center of Socialist Studies, Research and Education). CERES was associated with the PS, or Parti socialiste (Socialist Party), as is SOS-Racisme, which is an anti-racist Non-Governmental Organisation.

16 Augustin Cochin (1876-1916) was a French historian of the French Revolution who was killed in the First World War. Several of his works have been translated.

17 *Libération* is a daily newspaper with a Leftist viewpoint.

18 Alain Madelin (1946-), at the time that Faye was writing, was a member of the National Assembly of France and the President of the Démocratie Libérale (Liberal Democracy) party. He was known for his pro-American and laissez-faire economic positions. He retired from politics in 2007.

19 Alain Juppé (1945-), at the time that Faye was writing, was a Right-wing member of the National Assembly of France. He was Prime Minister of France under Jacques Chirac from 1995 to 1997. Following a conviction for the mishandling of public funds in 2004, Juppé is presently only the Mayor of Bourdeaux.

The reason for this? Intellectual movements that gain public attention *pose challenging problems along with a concrete political threat.*

The Nouvelle Droite has thus found itself in an increasingly precarious situation, devoid of any political backing and cut off from its own natural public, whose outlook was for the most part close to that of the Front National. The 'public of the Nouvelle Droite' was puzzled by our Third-Worldist and pro-Islamic positions, which were ideologically incomprehensible and were taken as the expression of a 'bourgeois way of thinking' indifferent towards immigration problems, or even as evidence of flirtation on our part with the non-Jacobin Left. From that moment onwards, unable to appeal to a new public, the Nouvelle Droite was progressively enveloped by the Front – the cultural value of its publications simply could not make up for its ideological drift. No doubt, as we shall see, increasing hostility on the part of the media also made the spread of the ideas of the Nouvelle Droite increasingly difficult. Like Ruyer and Freud (but not Debord, a rehabilitated para-Marxist), de Benoist's work has been confined by the system to limited spheres of influence.

But make no mistake: this is no excuse. The strong pressure put upon decision-makers by well integrated minority circles and lobbies such as SOS-Racisme, MRAP, LICRA, DAL, Ras l'Front, LDH, ACT-UP or Greenpeace and the various ideologues that inspire them cannot exclusively be explained on the basis of their political ultra-correctness and total complicity with the system: it is also due to the fact that they have been capable of powerfully delivering their messages, by using all the tricks of the new media circus. The Nouvelle Droite has not managed to do the same, but has remained tied to an obsolete view of how ideas are circulated.

The surfacing within the European population of a persistent faction destabilised by the 'crisis' and revolting against the concrete results of the system would have served as a new breeding ground for the Nouvelle Droite.

2. The Tightening of Censorship and the Nouvelle Droite's Failure to React to It

In the early 1980s, soft totalitarianism against all 'incorrect' forms of expression tightened its grasp. Once the generation of '68 – which used the slogan 'It's forbidden to forbid' – came into power, it distinguished itself for its conformism, taste for prohibition and desire for *ideological order*.

Censorship is exercised both through the legislative erosion of free thinking and writing (even by the use of lawsuits) and – in most cases – through intentional silence on the part of the media when it comes to people or things that might prove upsetting: a demonising and blacking out strategy. The Nouvelle Droite has certainly fallen victim to this censorship, which was even made the object of a GRECE meeting. But let us not exaggerate things. I fear that censorship is invoked as a pretext to justify lack of will and the failure to take any risks.

Each form of censorship represents a stimulus, each form of oppression a challenge: one should rise and face challenges, not complain. Why, was the Nouvelle Droite threatened with a ban? With persecutions and violence? Truth is, it was never capable of managing and turning to its own advantage 'conformity of ideas' (a pertinent notion first coined by Alain de Benoist and later mediatised by Jean-François Kahn,[20] who – paradoxically – is actually a lackey of political correctness and hegemonic thinking).

On the other hand, at the height of its glory – starting from 1979 – the Nouvelle Droite was subjected to a number of serious media attacks and even physical acts of aggression, but it was this very air of battle that had given it its drive and elicited creative reactions on its part.

There is no need to come up with bad excuses, overemphasising the treachery and effectiveness of censorship. Silence on the part of the media can also be explained with their *indifference towards the Nouvelle Droite*, a movement which is no longer surprising, shocking or provoking; which, despite the clear value of its writings, has ceased to offer anything new. I bet – and will get back to this point – that had the Nouvelle Droite resumed its former fighting spirit, had it sought to launch provoking debates and formulate *radical ideas*, the media black-out would have been short lived: the media must necessarily attack – and hence advertise – everything that opposes their system. I am paid to know it.

Aggressions are opportunities: they mediatise thought and enable it to grow sharper and react. With both skill and daring, one must outrage people if he wishes to be listened to; and most importantly, he must prevent his thought from becoming bourgeois.

---

20 Kahn (1938-) is a French journalist known for his liberal viewpoints.

3. An Incorrect Publishing Policy

The Nouvelle Droite has three magazines (which are more like light buoys than lighthouses): *Nouvelle École*, *Krisis* and *Éléments*. The function of the first two of these, which are theoretical in nature, is to establish intellectual foundations. By contrast, *Éléments*, the chief media bridge, is badly positioned: it is intended to be the cutting edge among Nouvelle Droite publications, to address an educated public and persuade people in outside milieus, but it falls short of its target. It lacks dynamism, addresses too many literary and intellectual topics that do not serve its purpose, and engages with few social issues; it contains long, stiff and often repetitive articles, and inadequate graphics with bad captions – defects that limit its media appeal. The layout of the magazine, particularly in its new version, is aesthetically impeccable, yet too austere and quite unsuited for an ambitious publication.

Still, behind all this, talent is still to be found. Editorial blunders alternate with excellent reports, although there are too few of the latter. The enquiry on the noxiousness of cars and the dead end reached by 'progress', for instance, which are featured in issue 86 (October 1996, 'La société folle') represents an example of what *Éléments* should be discussing systematically: topics of great interest to everyone and which capture readers' attention – a sort of intellectual detoxification and ideological revival.

While the 'analyses' made are often very sharp, and concrete, practical *theses* and suggestions are lacking that go beyond mere criticism and raise questions such as, 'Let's open the debate: what is to be done?'

Another mistake is *publishing dispersion*. I first noticed this shortcoming in the early 1980s. We should not multiply our publications, but concentrate our forces.

Charles Champetier introduced me to the small magazine *Cartouches*, which is full of dynamic and stimulating invective. Fine, but... Anyone working in the communications sector could tell you that the logic behind this magazine should be incorporated (and merged with) *Éléments*. Short pieces, striking information, a style that isn't stuffy, etc. Even *Krisis*, a magazine regarded as 'presentable' – but why? – tends to overlap with *Nouvelle École* and all too often succumbs to the appeal of Parisian slang, which does not always help carry on the debate...

To sum up my argument: I believe that some texts can only be aimed at 'inner' circulation, but that many others can be presented and circulated 'outside', at the heart of the system. We should never underestimate our own skills: talent always prevails over censorship, when it is accompanied by daring and intelligence.

Ideological Mistakes

The ambiguity of the ideological line of the Nouvelle Droite, which became more marked in the 1980s, constitutes the chief reason for its decline. To this, despite high-quality analytical texts – I am thinking, for instance of Champetier's work *Homo Consumans* or Alain de Benoist's article on 'colours' in issue 50 of *Nouvelle École* – we should add a return to doctrinal invective and a sort of intellectual bombast.

Let us now examine these mistakes.

1 – From the start, the members of the Nouvelle Droite and GRECE – including myself – have practiced *semantic clumsiness* and *permanent slips*. The double talk of many articles, magazines and books was caught between oblique references to issues, authors and iconographic motifs typical of the far Right – particularly that of Germany – and anti-racist, pro-Islamic, pseudo-Leftist or Third-Worldist tirades which did not fool the enemy, but puzzled our readership. I am happy to point to these shortcomings for which I too was responsible prior to realising how noxious they were. Today the Nouvelle Droite has not rectified these mistakes, but if anything worsened them.

2 – Second serious mistake: *the exploitation and politicisation of paganism*. Starting from a correct Nietzschean assessment – regarding the egalitarian, levelling and ethno-masochistic harmfulness of Christian evangelism – the Nouvelle Droite has set up a neo-pagan corpus that suffers a number of handicaps. Paradoxically, the unconscious starting point of this neo-paganism was a Christian perspective: the countering of dogma with a counter-doctrine. Paganism, as such, is non-existent: what exist are different, potentially countless kinds of paganism. The Nouvelle Droite presented itself as a 'pagan Church', one – moreover – without any deity. But paganism, by its very nature, is

unserviceable as a metapolitical banner, unlike Christianity, Islam or Judaism.

Second handicap: a virulent anti-Catholicism (where indifference would have been more in order), at times bordering on anticlericalism, combined with an open friendliness towards Islam. The latter is a risky attitude, given that Europe is facing a concrete Islamic threat, and represents a particularly absurd ideological stance, considering that Islam is a rigid theocratic monotheism, 'religion of the desert' in its coarsest form – far more than classic Catholic henotheism, which is strongly mixed with pagan polytheism. Moreover, the essence of the pagan perspective is to position oneself not 'against', but 'after' or 'alongside' – something which strikes me as being far more creative and innovative. I personally adopted this mistaken approach, which the Nouvelle Droite never corrected.

Third handicap: this paganism was – and still seems to be – marked by forms of folklore that find no space in the actual culture of Europeans (as opposed to what takes place in the United States!), and which I have always struggled against in a friendly spirit but in vain.

The result of this: one potential public never turned towards the Nouvelle Droite, while another fled from it. Why? Firstly, because many people could not understand this *preference assigned* to paganism over more important and concrete political matters, such as the destruction of the European ethno-sphere and anti-natalist masochism on the part of governments. Another consequence: the media effect of the promotion of paganism as a brand name, particularly in France, was to stir repulsion. An explicit appeal to paganism 'gives the idea of a sect', as I was once told by a great French actress, who was privately close to the ideas of the Nouvelle Droite, but unwilling, like many others, to mingle political ideology with para-religious elements. Such an attitude may be deplored, but still: there are certain rules of propaganda that cannot be ignored.

As for the attacks against the Catholic Church, these would have been – and would be – more apt if directed against the para-Trotskyism, immigrationism and self-ethnophobia of the high clergy, which favours a return to the hard evangelism of the original monotheistic sources, the 'bolshevism of Antiquity'. This masochistic and stupid high clergy that with false contrition is favouring the erection of mosques on European soil!

AN ASSESSMENT OF THE NOUVELLE DROITE

Two books have contributed to shape my outlook: *The Antichrist* by Nietzsche[21] and *The Gods of Greece* by Walter Otto.[22] As did Pierre Vial's initiatory 'oath of Delphi' in the early 1980s. By Apollo's sanctuaries, at sunrise, followers from Greece and Burgundy, Tuscany and Bavaria, Brittany and Wallonia, Flanders and Catalonia swore to keep the *pagan soul* alive. That's all very well, but pagan actions such as this should remain inside affairs.

The pagan soul is an inner strength that must permeate all ideological and cultural expressions. It is like the heart of a nuclear reactor: it is not something to be openly displayed through instrumental slogans. One doesn't go around saying 'I am pagan'! One *is* pagan.

More prosaically, I believe that this insistence on paganism as a para-political banner has puzzled the natural public of the Nouvelle Droite, as if the wish were to divert attention from secondary matters, while also starting an artificial conflict with 'traditional Catholics', who are not all that Christian after all... The exploitation of paganism has been a huge communications and propaganda mistake, which has distanced the Nouvelle Droite from many Catholic milieus initially favourable to it, which shared its ideas but were sentimentally tied to local traditions. We have made this serious blunder from the start, and it still waits to be rectified.

3 – Third mistake: an overemphasised *folklorism* and excessive cult of *rootedness*. The soul of European artistic culture lies not in small pyramidal objects of baked clay, painted furniture from Schleswig-Holstein, Breton bonnets or the naïve wooden sculptures of Scandinavian farmers; rather, it is found in the Reims cathedral, the double-helix Italian stairway in the Château de Chambord, the drawings by Leonardo da Vinci, the comics by Liberatore and the Brussels school, the design of Ferraris and the German-French-Scandinavian Ariane 5 rockets. By reducing European culture to mere folklore, this is depreciated and dragged down to the level of the 'primitive art' so

---

21   *The Antichrist* was one of Nietzsche's last books, written in 1888. In it, he attacks the problems of modernity, which he saw as being rooted in the deficiencies of Christian theology, which he challenges in this book. Famously, he calls for a 'transvaluation of all values'. Many English translations exist.

22   Walter Friedrich Otto (1874-1958) was a German classical philologist who was the administrator of the Nietzsche Archive during the Third Reich. Some Christian theologians have attacked Otto for attempting to revive ancient Greek religion, although he himself dismissed such notions as absurd. The book Faye mentions has been translated as *The Homeric Gods* (London: Thames and Hudson, 1979).

dear to Jacques Chirac. What ought to have been done, with Nietzschean anti-egalitarian logic and Cartesian 'common sense', was affirm the superiority – that's right: the *superiority* – of European artistic and cultural forms above all others. But the *ethno-pluralist* dogma – which stands in contradiction to *anti-egalitarianism* – prevented this. Having put too much faith in ethno-cultural relativism, and imbued with the guilt-stirring masochism that is so widespread, we didn't dare affirm the superiority of our own civilisation. Had we carefully done so, we would have appealed to a wide public of people who would have been struck by our daring.

Too many writings on European 'traditions', often connected to defunct or mythical folk customs, made us forget the crux of the debate: the *self-affirmation* of contemporary European culture, the geo-demographical threats looming over it and the need for a *reconquista*.[23] Folklorism, acting as a levelling mechanism, has situated European culture on the same level as others, when it was instead necessary to implicitly and adroitly affirm its creative primacy. On the other hand, this often folkish traditionalism serves the conquering spirit of American 'cultural products': it neutralises European culture and renders it into museum exhibits. Folklorism has failed as an identitarian bond for the contemporary cultural battle, and is having a disarming effect instead.

Contemporary European culture is creatively resisting in many fields: music, architecture, design, leading technologies, sculpture... The Nouvelle Droite has not paid adequate attention to this.

4 – The fourth mistake lies in the insufficient attention paid to *concrete problems*. The Nouvelle Droite, today even more than in the past, is too concerned with what may be termed *culturalism* and *historicism*. In the late 1970s it had achieved a degree of mediatisation and influence thanks to its ideological inroads and new debates on eugenics, the biological revolution, I.Q. differences among various populations, ethnology, new economic perspectives, the place of sexuality in the society of spectacle, etc. I believe that the Nouvelle Droite and its publications verge too much on the side of commemoration, literary culture, and

---

23 *Reconquista* is a Spanish word meaning reconquering or recapturing. Historically, it refers to the struggle of the Christian Spaniards against the occupation of Spain by the Muslims during the Middle Ages. The present-day Right uses it to refer to the reclaiming of European lands from non-European immigrants.

AN ASSESSMENT OF THE NOUVELLE DROITE

antiquated, nostalgic forms of intellectualism. This is a real shame, for the few treatments it gives of crucial contemporary issues are of high quality, as can easily be discerned from the pages of *Krisis*.

I wouldn't like to give the wrong impression: I am criticising the Nouvelle Droite not so much for what it does, as for that which it *does not do or no longer does* – or, to be objective, for what it does not do enough.

It is necessary to discuss things such as the Asian financial crisis and the biotechnology revolution, and launch discussions and debates on issues such as European federalism (for or against the United States of Europe?), the effects of the Internet, European space policy, the solar system, the deterioration of the environment, the consequences of an ageing population on pension funds, the boom in Latin American music, the outburst in female homosexuality, the world of pornography, sport, the demographic colonisation of Europe, energy policies and nuclear energy, transport and crime.

The Nouvelle Droite will only prove creative and credible once more if it manages to *formulate disorienting doctrines* regarding all major contemporary issues and if it is able to establish a new ideological corpus – presented in the form of a 'debate' rather than dogma – on economic, scientific, geopolitical and sociological matters.

5 – Fifth ideological mistake: Third-Worldism. I have fully contributed to this and am willing to exercise self-criticism. Alain de Benoist's essay *Europe-Tiers-monde, même combat*[24], a crucial work on the matter, and the articles I myself wrote on the issue in the early 1980s, driven by misdirected anti-Americanism, have been ideological and strategic impasses which have worried me since. No folk in history fights 'the same battle' as other peoples: every alliance is temporary. Besides, the very notion of 'Third World' has crumbled. What we have are China, India, the future Muslim Empire… The 'Third World' does not exist. Third-Worldism (which in our political milieu served as an awkward certificate of anti-racism) ignores actual history: the immigration and geopolitical pressure of the South against the North. What is worse, this misplaced Third-Worldism has been accompanied by a disconcerting and naïve pro-Islamic stance to which we all succumbed when an objective, aggressive, revanchist and comprehensible threat

---

24 Or *Europe-Third World: The Same Struggle* (Paris: Laffont, 1986).

was actually being posed by the Arab-Muslim world against Europe, seen as a 'land to be conquered'. It is quite true that dogmas make you blind. They are also dangerous: it is clear that for the most part the public of the Nouvelle Droite – and others too – did not share these surrealist views of ours.

6 – Sixth ideological mistake: *anti-Americanism and the feeling of being colonised*. In the early 1970s, in line with the anticommunism that was still prevalent on the Right, GRECE was pro-American and supported the 'West'. Thus in an old issue of *Nouvelle École*, under a photo of the Rockefeller Centre in New York, we find the following caption: 'The energy at the heart of power.' In 1975, however, Giorgio Locchi made us do an about-face: a special issue of *Nouvelle École* was published by Alain de Benoist and Locchi, which divided the civilisation of the United States from that of Europe, its roots. Later on, following the same drive, I suggested an alternative ideological axis, based on the separation of Europe from the West – a revolutionary idea in a milieu which made the 'West' its banner. We sought to affirm the idea that the notion of 'Western civilisation' or 'Western ideology' was not necessarily compatible with the destiny of Europe as a *land of brother peoples*. Western – the 'West' – is an abstract geographical notion, while the true fracture is between North and South: for the vital geopolitical space of Europe extends out to the Russian Far East. This was the ideological axis.

It was distorted, however, by the mistaken assumption that a structural solidarity exists between the peoples of Europe and those of Africa, Asia and Latin America against the Yankees. Actually, as we shall see, the United States is better regarded as a rival and opponent (*inimici*) than as an enemy (*hostes*).

7 – Seventh mistake, no doubt the most serious of all: the ambiguity of the catchword *ethno-pluralism*, which is worsened today by the addition of the predicate *multiculturalism* and by *inter-ethnic* communitarianism. These have been adopted by the Nouvelle Droite and I regard them as complete ideological impasses.

Ethno-pluralism initially possessed an implicitly 'external' meaning: all peoples are different and should be respected, yet each should live in its own land, in a well-defined ethno-cultural sphere, while cooperating with others. This implied a rejection of migration flows towards Europe and of the idea of a global ethno-cultural melting pot (actually, only Europe is the destination of these migrations). So far, so good: this is a consistent view. But the Nouvelle Droite – see, for instance, issue 91 of *Éléments*, published in March 1998, and which refers to the 'challenge of multiculturalism' on its front page – sought to give the notions of ethno-pluralism and multiculturalism an 'inner' meaning that stands in contrast to the first, for instance by vehemently defending the use of the Islamic veil in schools. By acknowledging the presence of separate ethnic communities on European soil, it turns ethno-pluralism into the vehicle for a tribal, ghettoised (and perfectly American) view of our society, which stands in contrast to the very meaning of the expression 'each folk in its own land'. Ethno-pluralism has thus been distorted in such a way as to deny the notion of European folk and even of 'folk' in general. Here too, the public is lost: similar stances puzzle our natural readership, while failing to convince our enemy that we are politically correct.

My criticism towards the ethno-pluralism and multiculturalism of the Nouvelle Droite can be summed up as follows:

Firstly, the Nouvelle Droite minimises – either for altruism or ignorance of ethnic and socio-economic events – the *catastrophe* represented by demographic-shifting immigration into Europe, a land which, unlike the United States, is generally only adapted to intra-European movements. There are three aspects to this catastrophe: rapid ethno-anthropological alteration; the erosion of European cultural roots (for which Americanism is less to blame); and strong economic and social setback, leading to poverty and endemic crime. The contemporary communitarian and multiculturalist discourse of the Nouvelle Droite can be interpreted as a sort of *fatalism*: for it sees the ethnic kaleidoscope of Europe, multiracial society and immigration as ineluctable events we should accept and submit to, managing and putting up with them as best we can. This is a demobilising stance, which is incompatible with an ideology that regards itself as revolutionary – although ultimately it proves to be 'politically correct'.

It is a sign of weakness to justify multiculturalism by invoking globalisation and the decline of the nation-state (which are self-evident facts). Only Europe and the United States are being made the victims of demographic colonisation from the South. But while the United States can withstand it, Europe cannot. All across the world, what we are witnessing is the *self-affirmation of vast, homogeneous ethnic blocs*, not multiracial 'communitarianism'. The prospect of a 'multicultural' planet is a Disneyland dream, a peace-lover's error. *The future belongs to peoples*, not tribes. The Twenty-first century will witness global ethnic warfare and the legions of immigrants in Europe will serve as the 'fifth column' of an aggressive South. This is not paranoia: it is geopolitics. To walk or drag one's feet in the footsteps of the blinding, immigrationist pacifism of European Leftist intellectuals is to make a serious mistake that threatens to soon lead the Nouvelle Droite to its ruin.

Accusations of 'paranoid rhetoric' against those who fear the immigrant 'invasion', 'Islamisation', fundamentalism and 'ethnic war', and believing that the repeated revolts in the *banlieues*[25] are the work only of alienated and Americanised youths with no roots (who could be perfectly integrated, *if treated nicely*) derives from a serious error of judgment, caused by an abstract ideology that ignores social events. The ethnic war in France has already started. The barbarisation of society and the rancorous and latent aggressiveness towards European culture shown by a large portion of young people brought here by immigration constitute an intermediate-term threat, as many impartial American sociologists have also observed. Why not acknowledge this?

On the other hand, the Nouvelle Droite envisages a model of social harmony within a pacified multicultural society, which is sheer utopia. Every multiracial – and multicultural – society is multiracist and 'infra-xenophobic': from Brazil and former Yugoslavia to Algeria, Black Africa, and the Caucasus. Multi-ethnicism in France will prove explosive and will have nothing to do with the placid tribalism my friends Alain de Benoist and Charles Champetier have outlined (see issue 50 of *Eléments*) via a discourse that may be taken as an example of 'the sociology of dreams'. Tribalism is never peaceful. I am ready to bet that, within ten years, history – through painful experiences – will have made all multiculturalist plans unserviceable, *even for those on the Left*. Alain de Benoist's wish is to 'foster a fruitful exchange of dialogue between groups that are clearly situated in relation to one another' (*Eléments*, issue 50, p. 3). This, in European soil, strikes me as a rather unfeasible prospect, which derives from the same ideological illusion that inspired the advocates of 'ethnic harmony' in 1950s

---

25 *Banlieues* means suburbs. Unlike in other countries, however, Parisian suburbs are associated with low-income housing for immigrants, making them more similar to British 'Housing Estates' or American 'projects'.

America, who opposed the idea of the assimilating melting pot. Actually, I believe that both assimilators – Jacobins and people in favour of the melting pot – and communitarians are wrong. A society based on ethno-territorial co-existence was, is and will always be impossible. *One land, one people*: this is what human nature requires.

I completely agree with the anti-Jacobinism, organicism and polycentric social view promoted by my aforementioned friends. What I reproach them for is their failure to admit that the harmonious socio-cultural diversity they are talking about can be achieved *only among different but related European peoples*. Out-and-out Europeanists, why do they believe or pretend to believe that a harmonious society will be established in France through 'multicultural' cohabitation with communities of Asian, African and Arab-Muslim origin, which are far removed from the mental framework of Europeans? Were they really consistent, they would defend the hard and abstract Republican idea of forced integration dear to Madame Badinter.[26] In this respect, the 'harmonicism' of the Nouvelle Droite is self-contradictory. They insist on promoting a paradigm that is *physically* impossible to implement, submitting to the faith in miracles that characterises egalitarian ideologies.

My friends of the Nouvelle Droite have an imaginary view of Islam. They believe Islam can be integrated within a model of European harmony and general tolerance, without taking account of the fact that this ultra-monotheism is an intrinsically conquering, theocratic and antidemocratic religion that seeks – as General De Gaulle had foreseen – to replace each church with a mosque. By its very nature, Islam is intolerant, exclusivist, and anti-organic. The current thinkers of the Nouvelle Droite are captivated by the senseless talk about 'French Islam', and fail to realise that they are facing the *strategy of the fox* Machiavelli[27] so aptly described. While followers of Carl Schmitt, *in practice* they never apply either the concept of the 'exceptional case' (*Ernstfall*) or that of the objective enemy: *he who identifies you as an enemy for the very reason you exist, whatever you may do*.

---

26 Élisabeth Badinter (1944-) is a prominent French feminist philosopher who also advocates the abolishment of cultural differences between populations in France, feeling that they only generate conflict.

27 In Chapter 18 of *The Prince*, Machiavelli writes: 'A prince, therefore, being compelled knowingly to adopt the beast, ought to choose the fox and the lion; because the lion cannot defend himself against snares and the fox cannot defend himself against wolves. Therefore, it is necessary to be a fox to discover the snares and a lion to terrify the wolves.' From the translation by W. K. Marriott (London: Dent, 1911), pp. 137-138.

The multiculturalism and pro-Islamic stance of the Nouvelle Droite are objectively close to the incautious positions adopted by the Catholic episcopate in France, which also believes – out of altruism – in the idea of a future harmonious and ethno-pluralist society on European soil.

Stranger still is the fact that the Nouvelle Droite does not seem to realise that in Muslim eyes 'pagans' are absolute enemies and spawns of the devil, while they are instead tolerated – even if looked down upon – by Jews and Christians. In a recent trip of mine to Saudi Arabia, I had to write 'Catholic' on the identity card given to me on board the plane: had I written I was a 'pagan' or follower of any other non-monotheistic religion, I would have faced some problems.

To expect an agreement between paganism and Islam is like hoping to reconcile the devil with holy water.

In its report on multicultural society, Éléments does not discuss the issue of the *impossibility of expelling illegal immigrants* (on account of reactions on the part of para-Trotskyist associations and Leftist Christians); nor does it discuss the social and economic cost of immigration, or the ongoing arrival in Europe of immigrants from the South: are we to seal this breach, and if so, in what way? Crucial questions such as this are never raised: yet people are waiting. There is also another problem: while each year tens of thousands French graduates leave for the United States, France is welcoming – and in exchange for what? – tens of thousands immigrants from the South with no qualification. Why not discuss this? Because it's taboo? That's right.

I reproach the Nouvelle Droite for its adherence to a worldview that is undermined by a devastating concept: 'realism' – which often takes the form of disheartened fatalism.

I am Nietzschean and do not like the term 'realist'. History is not realist. Communism collapsed within three years: who would realistically have foreseen that? In issue 5 of Pierre Vial's magazine *Terre et Peuple*, historian Philippe Conrad[28] illustrates the Spanish *reconquista* against the Afro-Muslim invaders, emphasising that in history there are no 'accomplished facts'. The *reconquista* was an unrealistic yet concrete endeavour, and it was accomplished. *The essence of history is both real and unrealistic*, for its motor is comprised of both fuel – *will to power* – and combustive – *the power of will*. Those who out of weakness choose to give in when faced by disagreeable and coercive

---

28 Philippe Conrad (1945-) was a historian and a member of GRECE.

AN ASSESSMENT OF THE NOUVELLE DROITE

historical events should heed the words of William of Orange: 'Where there's a will, there's a way.'

The mission of the Nouvelle Droite ought to have been to anticipate and pave the way for this path. It needs to correct its mistakes, by allying itself with other groups in Europe that agree with the above analyses.

The most effective ideological line would seem to lie in *simultaneously* rejecting multicultural and multiracial society on the one hand, and the Republican, Jacobin French nationalism that encourages it on the other. Yes to a great federal Europe; no to a multicultural (and in practice multiracial) France and Europe open to increasingly numerous Afro-Asiatic and Muslim communities.

8 – Eighth and final ideological gap: *the lack of an economic doctrine*. I had once started suggesting an economic doctrine for the Nouvelle Droite, one centred on the notions of 'organic economy' and 'autarchy for wide areas', as well as on a 'political' – as opposed to economic and fiscal – understanding of public authority. This doctrine called for the self-sufficiency of the great global power blocs, including Europe and later Euro-Siberia, with internal free exchange. This sort of thinking – which is compatible with the building of Europe – needed and still needs to be further developed.

Why? Because, as Henning Eichberg[29] had grasped – during a conversation between the two of us in Nice (in 1973!) – in order to change public opinion, and influence the course of history, it is necessary to 'talk about things' and not merely of 'abstract ideas': things that interest people. Spiritualism is necessary to give the movement a soul, but is not enough in itself. It is necessary to measure oneself with the eternal materialism of men. Like Marx (unfortunately), I believe that economy is part of the infrastructure of human concerns. In order to re-establish an effective ideological corpus, it is essential to possess an alternative economic doctrine. This means a return to concrete problems and social issues that affect people's lives: urbanism, transport, fiscal policy, the environment, energy policy, health care, birth rates, immigration, crimes, technology, television, etc.

* * *

---

29 Henning Eichberg (1942-) is a German sociologist and historian who has long been active in the German Right, and also founded the German branch of the New Right in 1970.

Of course, all these remarks on the ideological mistakes of the Nouvelle Droite do not mean that I am suggesting the adoption of a dogmatic ideological line. Simply, I believe that its 'official' doctrine is an impasse and that if it continues to be voiced it should be counterbalanced. A rather straightforward way for the Nouvelle Droite to regain credibility would be to *launch debates*. The issue of *Éléments* on multiculturalism – a central problem – would have drawn more attention, had it been open to contrasting opinions. For the magazines and events of the Nouvelle Droite to regain strength, they should follow this strategy: first, raise crucial and politically incorrect problems; and two, elicit debate on various sides.

I believe the Nouvelle Droite has lost influence because of its establishment of ambiguous and incomprehensible ideological axes. Its members have been too close to academia, too sophisticated and too charmed by quasi-Leftist, pacifistic, utopian and integrationist debates. It is necessary instead to take a resolute stance and *make a clear break with the system by developing a radical and revolutionary school of thought*. We should be wary of false wisdom and false friends, of false acknowledgements, successes and – most importantly – *false good ideas*. Wrong ideas have the seductive elegance of decadence, not the 'modest and simple harshness of truth' (Nietzsche). An ideology can only prevail by setting itself in opposition to an already declining order.

The Nouvelle Droite – and mine is a very friendly exhortation – should draw new energies from Nietzsche's 'philosophy of the hammer'.

The Nouvelle Droite, or those who will take its place on the spectrum of ideologies in Europe, will only prove successful through the virtue of courage. If *through the art of discussion and without any dogmas, they are able to develop a radical and politically incorrect thought*, even through the use of current *forms* of expression and communication.

The Nouvelle Droite has not been a 'victim of the System' or of 'censorship', but of itself. Nothing is lost for those capable of rising again.

For today, as my friend Giorgio Locchi foresaw, we are entering the dark age of storms, the *interregnum*: a century of battle and steel, decisive for the future of the European peoples and their offspring – an age that calls for a *tragic and combative ideology*.

It is necessary for organisations to formulate efficient and dynamic paradigmatic ideas: original and daring ideas capable, like weapons, of averting incumbent threats. The exponents of our current of European thought must come together and adopt the *optimism of pessimism*: they must offer a will, an axis, for this great fatherland that is being built amid haze and pain. Like a sleepwalker driven by his self-assurance, half-conscious of the threats looming over him, in a chilly turmoil an empire is rising that does not dare yet to utter its own name; historical thunder that is being born amid the pains of childbirth: Greater Europe. Our only hope for survival.

An idea is well founded only if it conforms to a concrete historical perspective, only if it is the expression of a sincere hope.

New Ideological Paths

What follows are what I believe to be the axes and paths for an ideological regeneration, which I will define in more detail further on. Here are a few suggestions:

1 – First is what I would term *vitalist constructivism*, an overall ideological framework that unites an organic and daring approach to life with the complementary worldviews of Nietzschean will to power, Roman order and realist Hellenic wisdom. Leitmotiv: 'a concrete voluntaristic thought that creates order.'

2 – The second axis may be defined as *Archeofuturism*: to envisage a future society that combines techno-scientific progress with a return to the traditional answers that stretch back into the mists of time. This is perhaps the true face of post-modernity, as removed from attachment to the past as it is from the foolish cult of 'keeping up with progress': the harmonious union of the most ancient memory with the Faustian soul according to the logic of 'and' rather than 'or'. Intelligent traditionalism is the most powerful form of futurism – and vice versa. It is necessary to reconcile Evola and Marinetti, and do away with the notion of 'modernity' produced by Enlightenment ideology. The Ancients must be associated not with the Moderns but with the Futurists.

As the Nouvelle Droite noted, while the political and social structures of modernity are crumbling today, archaic ones are surfacing in all fields – a significant aspect of this phenomenon being the spread of Islam. Finally, the upheavals technological science –and particularly genetics – will cause in the future, like the tragic awakening to reality that is bound to take place in the Twenty-first century, will require a return to an archaic mentality. Modernism increasingly appears as a form of attachment to the past. Yet it is not a matter here of embracing classic 'traditionalism', which is tinged with folklore and yearns for a return to the past. Modernity has grown obsolete. The future must be 'archaic': neither modern nor attached to the past.

3 – Third axis: *to envisage the death throes of the European nation-state and European revolution* as the central political events of the Twenty-first century. This implies the need to jump on the wagon of unification if for no other reason than to correct its faults, even if – to use Lenin's words – it is *useful idiots* who are building the EU. Great revolutions never take place in a linear and vaunting way, as dogmatic and romantic intellectuals like to think. The painful gestation of the unification of the European *peoples* in their shared *land* – extending at first from Brest to the Oder and then from Brest to Bering – is an underlying movement that conceals an imperialist drive. This represents a reaction to decolonisation, the demographic crisis and immigration, and is possibly the solution to many problems that currently appear unsolvable. *Eurosiberia* is what we must now envisage. The assumption behind all this is the idea that the Earth, a global village and interdependent habitat, cannot be managed – particularly for environmental reasons – by a range of different national actors, but must rather be directed by a limited number of 'imperial blocs': Greater Europe, India, China, North America, Latin America, the Muslim world, Black Africa and peninsular Asia.

No doubt, this is a still distant scenario. Yet, the role of 'thinkers' is to foresee the future. Today we must launch the idea of a *United States of Europe*.

4 – Fourth axis: to think about the fact that in the Twenty-first century humanity will face a *convergence of catastrophes*. I will further develop this essential point later on. When forced with their backs against the wall, human societies always react. A series of macro-lines of catastrophe are converging towards a breaking point situated somewhere in the early Twenty-first century: an environmental, economic and

AN ASSESSMENT OF THE NOUVELLE DROITE

military apocalypse brought about by 'faith in miracles' – including the belief that 'development' can continue indefinitely without posing the risk of general collapse. The egalitarian civilisation sprung from modernity is now witnessing its last good days. We must now think about the aftermath of the catastrophe: we must already start *developing the vision of an Archeofuturist world for the aftermath of the chaos.*

5 – Fifth axis: to think about the conflict between North and South that is emerging – a possible third world war – and the role Islam may assume as the symbolic banner of revenge. This calls for a redefinition of the notions of the enemy and an objective threat: we must be wary of all erudite talk about the harmlessness of any 'global Islamic front' and tackle the issue of ethnicity, which may be added to the environmental and economic issues of the new iron century in the making...

In this view, we should stop always portraying southern countries, and particularly Africa, as the eternal 'victims' of the evil schemes of northern countries. The neo-colonialist martyr myth should come to an end. Each folk shapes its own destiny. We should have the courage to treat poor countries as responsible actors rather than victims: Africa's misfortunes are chiefly caused by Africans themselves – we cannot continue to beat our chests and act in their place. The Nouvelle Droite must distance itself from the *paternalistic post-colonial masochism* of the entire European intelligentsia, be it of the Left or the Right.

6 – Sixth axis, related to the first: is the United States an *enemy*, i.e., a potential destructive invader, or an *opponent*, i.e., a debilitating rival on the cultural and economic level? Does the United States – 'the only superpower for only twenty more years',[30] according to Zbigniew Brzeziński – really represent the chief enemy? Is it more dangerous than the South? I believe we are now closer to the Russians – our former absolute enemies – than we are to the Americans – our former absolute friends; yet, by *already* seeing ourselves as *Eurosiberians*, we must envisage a strategy of agreement or conflicting cooperation with

---

30 Brzeziński (1928-) was the National Security Advisor to the Carter administration from 1977-1981. Since then he has gained a reputation as a highly respected political analyst. I cannot identify this precise quotation, although in Brezezinski's 1997 book, *The Grand Chessboard*, he identifies the unique qualities and opportunities which have allowed the U.S. to become the lone superpower since the end of the Cold War, and predicts that given the new challenges the world faces and its increasingly multipolar nature, the U.S. will be unlikely to maintain this role for more than another generation.

America against a major threat from elsewhere. A clear break must be made with the myth of the United States as an 'invincible superpower': the U.S. is powerful because Europe is weak. It imposes nothing on us by force, unlike what the former USSR used to do with central European countries. The American imperial republic is right, from its point of view, to practice *soft imperialism*. We must gain control of our own destiny: we must be capable of distinguishing our mortal enemies from a rival opponent and, in any case, adopt a policy of self-affirmation.

7 – It is necessary to focus on the *epistemology of technique*. Problems: are not computer science and genetic engineering about to explode the framework of hegemonic egalitarian ideology, by creating an abyss between what is real and what is desirable, between nature and ultra-nature? These are crucial questions that concern biology and computer science. We must resume the debate we broke off concerning biology, as transgenic techniques today make it possible to intervene in the processes of genetic transmission which until recently were exclusively natural phenomena 'beyond the grasp' of all intervention. We are already capable of creating farm animals without gestation, in incubators, and shall soon be capable of doing the same with human beings: by combining advanced computer systems with transgenic techniques, we'll be able to program the gene pool and hence the abilities of 'second-generation humans'. From corn to sheep and from sheep to humans. An additional problem: third-generation computers will enable the creation of a virtual universe, or simulated anti-world, that will look more real than the real world, with genuine and autonomous hyper-virtual and three-dimensional characters, for 'computer intelligence' is dawning. Those who contemptuously claim that these 'are only machines' are making a serious mistake. These new blows against anthropocentrism, delivered by man himself, remind us that technological science is the Faustian spirit in action. Is this a deadly hazard for man, a 'diseased animal' and evolutionary failure? Or is it a destiny that can be governed? Such are the philosophical questions that every intellectual movement worthy of this name must address.

8 – It is necessary to reflect on the issue of immigration, which represents a form of *demographic colonisation* of Europe at the hands of mostly Afro-Asiatic peoples (and not an 'invasion', as put by the

demagogue Giscard,[31] the author of new regulations regarding family reunification).[32] Native Europeans are historically and objectively finding themselves is a situation not identical but very close to that of the American Indians and North African peoples in the Nineteenth century, when they witnessed the arrival of the European settlers who had left an over-populated continent. Three generations later, the colonisation of Europe represents a form of revenge against European colonisation. In organising a reaction it is necessary to *shift the centre of the debate*. This is not simply a cultural or socio-economic problem, as those discussing this issue would like to believe: it is a global anthropo-ethnic problem. It will be necessary to clearly emphasise this methodological distinction in the answer given (for or against) the real problem: are we to accept or reject a substantial alteration of the ethno-cultural substrate of Europe? The basis of intellectual honesty and the key to ideological success lie in the ability and courage to address the real problems, instead of attempting to avoid them.

9 – To envisage a two-tier global order, given the technological, social and environmental impossibility of extending the logic of 'progress and development' (i.e., 'faith in miracles') to the entire planet. Could we not imagine and foresee a scenario where most of humanity reverts to living in traditional societies that consume little energy, and are socially more stable and happy, while – in the context of globalisation – a minority continues to live according to the techno-industrial model? Might there be two parallel worlds in the future, the worlds of a new Middle Ages and of Hyperscience? Who would be living in each of these worlds, and in what numbers? All daring and creative thought must *think the unthinkable*. I believe that *Archeofuturism*, an explosive meeting of opposites, is the key to the future, simply because the paradigm of modernity is no longer viable on a global scale.

10 – In this perspective, it is necessary to reflect on the economic issue of *autarchy for wide spaces* (which may include Eurosiberia) and the

---

31 Valéry Giscard d'Estaing (1926-) was President of France from 1974 until 1981. He made a famous speech in September 1991 in which he referred to immigration as an invasion and called for tougher standards for aspiring citizens.

32 Family reunification is an immigration policy which allows for the entry of the family members of a foreigner who has become a citizen or a permanent resident of the country. This is upheld by the U.S., Canada and most Western European countries, and is one of the principal means for immigrants to legally enter those nations.

moving beyond both socialism and liberalism, by reviving the idea of an organic economy of the Third Way that may be inspired both by genuine liberalism and genuine communitarian socialism. We must think about the ongoing transformation of economic systems into neo-feudal networks, and radically redefine the role of superior political authority, which must *politically direct* the economy, but not manage it. We must envisage great semi-autarchic blocs which may have different modes of production and consumption, and within which interlinked but diverse types of society and economy may exist. Could ultra-technological areas, connected to the global communications network, border with neo-archaic areas where the ways of life and production of traditional societies have been restored?

A current of thought is strong if it manages to pose crucial yet unexpected questions, if it acts in advance – particularly if it does so with a non-dogmatic language.

* * *

In order for an ideology of revolution and restoration to emerge in this age of great challenges, where vital matters are at stake and catastrophes loom near, it is necessary to reformulate the old notion of *conservative revolution*, which I consider outdated. All the young forces, which are so few in these videophonic times, must unite on a European scale and forget about parochial disputes, hierarchically defining – according to the non-exclusivist and polytheistic logic of *and* – the *worldview* that unites them and the *doctrines* that launch the debate. *Ideology* will follow later. Finally, it would be necessary to balance critical discourse on this interregnum period with a precursory, assertive and optimist discourse within our pessimistic view of the present, which may apply to the aftermath of the chaos.

The keystone of our current of thought is an agreement – of an historical kind – on the notion of *Europe*. All of us – each according to his dreams, analyses, and temperament – wish to move beyond the obtuse nationalism of Enlightenment egalitarianism, and contribute to build this macro-continental union of brother-peoples, *preparing the idea of it for the aftermath of the catastrophe*. All this – in conformity with an *organic and democratic* imperial logic – without forcibly conforming ourselves to others and destroying the historical heritage represented by our various languages and ethno-cultural sensitivities,

AN ASSESSMENT OF THE NOUVELLE DROITE

which constitute Europe's unique treasure. These are the words of Pierre Vial, one of the leaders of the Front National – a French nationalist party – and the founder of the cultural association Terre et Peuple, 'This is the real purpose of our struggle: to fight for a rooted cultural identity which is both French and European, and which harmoniously combines the Greco-Roman, Celtic and Germanic heritages. Each of these is dear to us, for it is an aspect of one and the same civilisation. All those who are fighting to preserve this civilisation are our brothers in arms.'

We must become *soldiers of the Idea* again and in a flexible *yet* articulated manner federate all currents of thought, periodicals, books and associations following the same line on a European scale.

What struck me when I started reading publications from our 'movement' again – and this was only recently, because I hadn't been interested in such things for a while – was the existence in Italy, Germany, Belgium, France, Croatia, Spain, Great Britain, Russia, Portugal, etc. of men, magazines, movements, and associations that all adhere to a broadly similar worldview. But I was also struck by the dispersion, personal contrasts, and heated parochial spirit of some people.

A synergic movement of this kind that cuts across currents and tendencies, converging on the axial ideas I outlined above, will only manage to carve a place for itself in history if driven by provocative idealism rather than neutral *intellectualism*.

May my talented friends of the Nouvelle Droite benefit from these few words of advice to yet again find their path in history – perhaps they could start by changing their name...

2

A Subversive Idea: Archeofuturism as an Answer to the Catastrophe of Modernity and an Alternative to Traditionalism

To Giorgio Locchi and Olivier Carré, in memoriam.

1 – Method: 'Radical Thought'

Only **radical thought** is fruitful, for it is the only one capable of creating daring ideas to destroy the ruling ideological order and enable us to free ourselves from the vicious circle of a failing system of civilisation. To quote René Thom,[1] author of the catastrophe theory, only 'radical ideas' can make a system plunge into chaos – 'catastrophe' or a traumatic change of state – in such a way as to bring about a new order.

Radical thought is neither 'extremist' nor utopian, for if it were it would have no hold on reality; rather, it must anticipate the future by making a clear break with the irreparably worm-eaten present.

Is this thought *revolutionary*? It must be such today, for our civilisation is situated at the end of a cycle, not at the beginning of a new development, and because no school of thought has dared to be revolutionary since the final collapse of the Communist experiment. Only by outlining new perspectives on civilisation will it be possible to be harbingers of historicity and authenticity.

---

1 René Thom (1923-2002) was a French mathematician who made many achievements during his career, but is best remembered for his development of catastrophe theory. The theory is complex, but in essence it states that small alterations in the parameters of any system can cause large-scale and sudden changes to the system as a whole.

Why 'radical' thought? Because this goes to the *root* of things, 'to the bone': it questions the very worldview on which the present civilisation rests, egalitarianism – a utopian and obstinate idea that with its inner contradictions is plunging humanity into barbarism and economic and environmental horror.

In order to shape history it is necessary to unleash *ideological storms* by attacking – as Nietzsche correctly observed – the *values* that form the framework and skeleton of the system. No one is doing so today, hence for the first time it is the *economic sphere* (TV, media, videos, cinema, the show business and entertainment industry) that holds the monopoly over the (re)production of values. This clearly leads to a ruling ideology devoid of any ideas and creative, challenging projects: one founded instead on dogmas and anathemas.

Only radical thought today could enable intellectual minorities to create a movement, shake the mammoth, and deliver an electroshock (or shocking ideas, *ideoshock*) to stir society and the current world order. Thought of this kind must necessarily be non-dogmatic and must constantly reposition itself ('the revolution within the revolution', the only correct insight of Maoism), thus protecting its radical character from the neurotic temptation of fixed ideas, dream-like phantoms, hypnotic utopias, extremist nostalgias and raving obsessions – risks that threaten every ideological perspective.

In order to act upon the world, all radical thought must develop a consistent and pragmatic ideological corpus with detachment and adaptive flexibility. Radical thought is first of all a *query*, not a doctrine. What it suggests must be declined in the 'what if?' rather than the 'must' form. Compromise must be abolished, along with the false wisdom of 'cautiousness', the rule of ignorant 'experts' and the paradoxical conservatism ('status quoism') of those who adore 'modernity' and believe it will endure forever.

One last characteristic of effective radical thought: the acceptance of *heterotelia*, which is to say the fact that ideas do not necessarily yield the expected results. Effective thought acknowledges its own *approximate* character.

One sails by sight, changing course depending on the wind, yet always knowing where he is going and what port he is trying to reach. Radical thought integrates the *risks* and *errors* inherent in all human activities. Its modesty is inspired by Cartesian doubt and constitutes

A SUBVERSIVE IDEA 55

the driving force that sets spirits in motion. There are no dogmas here, only the power of the imagination and a touch of amorality: creative tension towards a new morality.

Today, on the eve of the Twenty-first century, which announces itself as a century of *iron and fire* – a century of colossal stakes laden with mortal threats for Europe and humanity at large – as our contemporaries lie stupefied by soft ideology and the society of the spectacle in the midst of a deafening *ideological void*, radical thought may finally be formulated and even affirm itself through the envisioning of new and once *unthinkable* solutions.

The insights offered by Nietzsche, Evola, Heidegger, Carl Schmitt, Guy Debord and Alain Lefèbvre[2] regarding the *reversal of values* can finally be put into practice, as can Nietzsche's *philosophy with a hammer*. Our 'state of civilisation' is now ready, which it wasn't in the recent past: for in the Nineteenth and Twentieth centuries – the centuries of modernity – it was breeding the virus without yet suffering from the infection.

On the other hand, we must reject the pretext that radical thought would be 'persecuted' by the system. The system is foolish. Its censorship is as far from stringent as it is clumsy, striking only at mythic acts of provocation and ideological tactlessness.

Among the official and acknowledged members of the European intelligentsia, thought has been reduced to the level of media mundaneness and the rigmarole of egalitarian dogmas out of fear of breaking the laws of 'political correctness', lack of conceptual imagination and ignorance of what is really at stake in today's world.

European societies today are undergoing a crisis and are ready to be permeated by radical and resolute ways of thinking that promote revolutionary values and *total dissent of a pragmatic rather than utopian kind* toward the present global civilisation.

In the tragic world that is emerging, radical and ideologically effective thought must combine the virtues of *Cartesian classicism* (the principles of reason, actual possibility, permanent examination and critical voluntarism) and *romanticism* (a dazzling thought appealing to emotion and aesthetics, along with daring perspectives), in such a way as to unite the virtues of the idealist philosophy of affirmation

---
2   Alain Lefèbvre (1947-) is a French journalist who was one of the founding members of GRECE in 1968.

with the critical philosophy of negation through a *coincidentia oppositorium* (coincidence of opposites), as Marx and Nietzsche did with their methods based on the 'hermeneutics of suspicion' (i.e., the indictment of ruling ideas) and the 'positive reversal of values'.

Thought of this kind, which combines daring and pragmatism, intuitive forecasting and watchful realism, aesthetic creativity and the will to historical power, must be 'a concrete and voluntaristic way of thinking capable of creating order'.

2 – Conceptual Framework: The Notion of Vitalistic Constructivism

My teacher, Giorgio Locchi, identified *egalitarianism* as the central axis and driving force – from both an ethical and practical perspective – of Western modernity, which is now failing completely. Stimulated by his writings, within GRECE we provided a wide critical and historical description of this phenomenon. For the future we promoted the idea of *anti-egalitarianism*, but this term was not enough in itself. A leading idea cannot merely be defined by opposition to something else: it must be affirmative and meaningful in itself. But what is the content, the active principle of this virtual anti-egalitarianism? *Of what does anti-egalitarianism concretely consist?* This question was never answered at the time, yet it is only through a clear answer that mobilisation may come about.

Influenced by the works of Lefèbvre, Lyotard,[3] Debord, Derrida[4] and Foucault,[5] as well as by the writings of architects such as Portzamprac,

---

3   Jean-François Lyotard (1924-1998) was a French philosopher who, from the 1970s onward, was one of the primary expositors of postmodern philosophy. His ideas have had a huge impact on the fields of philosophy, cultural studies and literary theory. As with all postmodernists, Lyotard rejected the idea of any type of universal meaning, claiming that meaning only exists when it is created by individuals or by small groups using their own narratives to understand reality. Thus, efforts to understand all of human experience within the context of a universal ideology, such as Communism or Fascism, are doomed to failure since they attempt to impose one system of understanding upon the universe, when actually meaning is something unique to every individual and thus cannot be extrapolated to others' experience. Many of his works have been translated.

4   Jacques Derrida (1930-2004) was a French philosopher who is widely regarded as the most important of the postmodernist philosophers. His work has had an enormous impact on philosophy and literary theory since the 1970s. Most of his work is available in translation.

5   Michel Foucault (1926-1984) was an erudite French philosopher, historian and sociologist who has been associated with both structuralism and postmodernism, although he rejected both labels. He wrote not only on philosophical themes, but also on the subjects of insanity and its treatment, prisons, medicine, and the history of sexuality. He was openly homosexual and a sadomasochist who died of AIDS, and he supported extreme Leftist ideas. All of his major works have been translated.

A SUBVERSIVE IDEA

Nouvel and Paul Virilio,[6] I sought to illustrate the need for *post-modernity*. Here too, however, the Latin prefix 'post', like the Greek 'anti', does not define any content. To affirm that egalitarianism and modernity (a theory and a practice) are irrational is not enough. Again, it is necessary to *imagine, state and suggest what would be good*. Any critique of a notion is only meaningful if it is accompanied by a new and affirmative notion.

But if this is the case, what leading idea(s) should we envisage? Allow me to explain this through a short recollection.

Together with the late painter of genius Olivier Carré, in the course of a subversive radio programme (*Avant-Guerre!*),[7] I had come up with a science-fiction tale of dark humour about an imaginary Eurosiberian Empire (the 'Federation'), whose white-and-red checkered banner was reminiscent of the flag of Angoumois, the tiny province where I (like Mitterrand)[8] was born, as well as that of Croatia. In particular, we used the term *vitalistic constructivism* to describe the titanic doctrine at the basis of one of the giant companies of that bizarre empire (Typhoone), whose aim was to move the Earth into another orbit... Later, with the benefit of hindsight, I realised that this radio gag, which also inspired a comic story,[9] possibly resulted from a failed ideological act on my part – a *lapsus linguae ac scripti*.[10] After all, Surrealism and Situationism had always taught that 'subversive ideas can only come from the pleasure principle'[11] (Raoul Vaneigem),[12] and that it is mocking and 'eccentric' brainwaves that should lay the foundations. Alain de Benoist has taught us that a person's style conditions his ideas. After all, André Breton[13] had already observed that 'gravity lies in what does not appear serious.'

---

6   Paul Virilio (1932-) is a French philosopher who writes primarily about technology, as well as what the use of physical space tells us about the institutions that utilize it. Many of his works have been translated.
7   'Pre-war'.
8   François Mitterrand (1916-1996) was the President of France between 1981 and 1995. To date, he has been the only member of the Socialist Party to become President, and was also the longest-serving President of the Fifth Republic.
9   This story forms the sixth chapter of this book.
10 Latin: 'slip of the tongue and of writing'.
11 This quote appears in Vaneigem's best-known work, The Revolution of Everyday Life (London: Action Books, 1972), although the quote actually reads: 'Discipline and cohesion can only come from the pleasure principle.' Faye appears to be paraphrasing, since in its original context, Vaneigem is referring to the discipline and cohesion of revolutionaries.
12 Raoul Vaneigem (1934-) is a Belgian philosopher who has written many books on anarchist themes. He is best-known for being part of Debord's Situationist International during the 1960s.
13 André Breton (1896-1966) was the founder of the Surrealist art movement in the 1920s, and wrote its most important essays and treatises.

So by further exploring this intuitive concept, I discovered four truths:

1 – Words, as Foucault argues (in *Les mots et le choses)[14]* have a crucial importance. They constitute the foundation of concepts, which in turn represent the semantic impulse behind ideas and the driving force of actions. To state and describe is already to construct.

2 – As Italian Communists have realised, there is no need to derive semantic denominations or aesthetic symbols from old and historically failed ideologies. Even the label 'Conservative Revolution'[15] strikes me as being too neutral, dated and historicised, tied as it is to the 1920s. Blind faith of this sort does not mobilise and is inadequate for the new challenges. In conformity with the *active* tradition of European civilisation we must launch new catchwords on the chessboard of history. The essence of the style remains, but the form changes. Each leading idea must be *furious and metamorphic*.

3 – The term 'vitalistic constructivism' provides an overall worldview and concrete synergic plan linking two mental structures. Through 'constructivism' it stands for: historical and political will to power, an aesthetic project of civilisation-building, and the Faustian spirit. Through 'vitalism' it stands for: realism, an organic and non-mechanistic mentality, respect for life, self-discipline based on autonomous ethics, humanity (the opposite of 'humanitarianism'), and an engagement with bio-anthropological problems, including those of ethnic groups.

4 – *Vitalistic constructivism* is the label I suggest we use to positively define what used to be called – for want of a better term – *anti-egalitarianism*.

Besides, anti-egalitarianism only defined its project within the vague and purely descriptive conceptual framework of *postmodernity*. The label I suggest to describe the central ideological plan of vitalistic constructivism is *Archeofuturism*. This I shall outline further on.

---

14 *The Order of Things: An Archaeology of the Human Sciences* (New York: Pantheon, 1970).

15 The Conservative Revolution is a term first coined by Hugo von Hoffmansthal, which has come to designate a loose confederation of anti-liberal German thinkers who wrote during the Weimar Republic (1919-1933), although scholar Armin Mohler, in his classic study of the movement, has identified Conservative Revolutionary thinkers going back as far as the Nineteenth century. There are some who speak of a 'conservative revolution' in today's world, representing the same spirit as the Weimar movement, although it is more commonly used to designate the historical school of thought.

3 – Diagnosis: Modernity Leads to the Convergence of Catastrophes

In order to define the content of a possible form of Archeofuturism, I must summarise the central points in my critique of modernity. Sprung from secularised evangelism, Anglo-Saxon mercantilism and the individualistic philosophy of the Enlightenment, modernity has managed to carry out its global plan, based on economic individualism, the allegory of Progress, the cult of quantitative development, and the affirmation of abstract 'human rights'. Yet, this has been a pyrrhic victory, for the plan implemented by this worldview, which seeks to claim the Kingdom of the Earth for itself, has entered a crisis and will probably collapse at the beginning of the next century.[16] After all, the Tarpeian Rock is on the Capitoline Hill.[17]

For the first time in history, humanity is threatened by a ***convergence of catastrophes***.

A series of 'dramatic lines' are drawing near: like the tributaries of a river, and will converge in perfect unison at the breaking point (between 2010 and 2020), plunging the world into chaos. From this chaos – which will be extremely painful on a global scale – a new order can emerge based on a worldview, Archeofuturism, understood as *the idea for the world of the post-catastrophic age*.

Here is a brief outline of the nature of these lines of catastrophe:

1 – The first is the widespread **metastasis of the European social fabric**. The demographic colonisation of the northern hemisphere by peoples from the South is becoming an increasing problem – despite all the reassuring statements on the part of the media – and one fraught with explosive consequences, associated in particular with the collapse of the Churches in Europe, which has become a land of conquest for Islam; the failure of multiracial society, which is increasingly racist and neo-tribal; the progressive ethno-anthropological metamorphosis of Europe, a veritable historical disaster; the return of poverty in both East and West, and the slow but steady increase in crime and drug consumption; the ongoing disintegration of family structures; the decline of the educational system and the quality of school curricula; the disruption of the passing down of cultural knowledge and social

---

16 As Faye was writing in 1998, he is referring to the Twenty-first century.
17 The Tarpeian Rock was a cliff located near the site of the Roman Forum on Capitoline Hill in ancient Rome. During the days of the Roman Republic and later the Empire, dangerous criminals and the physically or mentally disabled were executed there by being thrown off the cliff.

disciplines (*barbarisation* and incompetence); the disappearance of folk culture and its replacement by the brutishness of masses rendered passive by audio-visual technology (Guy Débord took his own life because he had already foreseen all this in 1967[18] in his book *Society of the Spectacle*);[19] the progressive decay of cities and communities in favour of sprawling suburbs devoid of all transparency and coherence, where there is no law or safety; endemic urban revolts – like a rampant May of '68, only worse; and finally the disappearance of all civil authority in the countries of the former USSR, which are overwhelmed by economic crisis. Meanwhile, nation-states witness their own sovereignty decline and prove incapable of facing poverty, unemployment, crime, illegal immigration, and the growing power of mafias and the corruption of the political class; the creative and productive elites, hit by taxation and increasing economic control, are thus allured by the prospect of moving to America. An increasingly egotistical and savage society on the road to primitivism, paradoxically concealed and counterbalanced by the naive and pseudo-humanistic discourse of 'hegemonic morality': this is what is emerging, year after year, and will soon reach a breaking point.

2 – These causes of social collapse will be worsened by an increasingly serious **economic and demographic crisis**. After 2010 the number of people working will be insufficient to fund the pensioners of the 'grandpa-boom'. Europe will collapse under the weight of the elderly: countries with an ageing population will witness the slowing down and crippling of their economies, for increasing resources will have be used to pay for health care and the pensions of unproductive citizens; besides, ageing limits techno-economical dynamism. The egalitarian ideology of (old) modernity has prevented any serious engagement with these problems, as it is paralysed by two dogmas: anti-natalism (a form of *ethno-masochism*), which censures all attempts to voluntarily increase birth rates; and the egalitarian refusal to pass from a social security system based on redistribution to capitalisation (pension funds). The worst is yet to come. Unemployment and poverty will increase, while a small class working in the international marketplace and supported by a class of bureaucrats and office workers with secure

---

18 See Introduction, note 7 about Debord. The more widely accepted reason given for his suicide on 30 November 1994 is that he wanted to end the pain inflicted by a chronic illness he had contracted as a result of his alcoholism.

19 *The Society of the Spectacle* (New York: Zone Books, 1994).

A SUBVERSIVE IDEA

positions will live comfortably. Economic horror awaits us. By a perverse process, egalitarianism is engendering a society of socio-economic oppression, thus showing itself to be the opposite of *justice*, as understood in Platonist terms.[20] Even the socio-democratic welfare state founded on the myth of Progress will collapse, and with a greater crash than the Communist system. *Europe is turning into a Third World country*. Before us is a time of crisis, or rather: the crumbling of the foundations of the socio-economical structure that has taken the name of civilisation.

America – a vast continent that has witnessed the migration of pioneers, and is used to a culture of brutality and a conflict-driven system based on ethnic and economic ghettoes – appears less vulnerable than Europe and more capable of facing an imbalance, at least when it comes to social stability; it too, however, would not survive a global maelstrom.

3 – The third fracture line of modernity: **the chaos of the South**. By setting a process of industrialisation in motion that undermined their traditional cultures, the countries of the South, dazzled by a deceptive and uncertain economic growth, created social chaos in their own lands, and now this is growing steadily worse. The recent events in Indonesia are a warning sign.[21] By perspicaciously turning his back on his own ideological family, the Anglo-French businessman Jimmy Goldsmith[22] had perfectly analysed all this: the explosion of huge urban conglomerates (Lagos, Mexico, Rio de Janeiro, Kolkata, Kuala Lumpur...) turning into hellish jungles, the presence of both poverty

---

20 Plato discusses his concept of justice at length in *The Republic*. In the context of the state, Plato saw justice in the ideal 'good city' as something that could only be attained in a state that was ruled by philosopher-kings, since he believed that philosophers' knowledge of Truth, and belief in Truth over self-interest, makes them less vulnerable to the corruptions of power than other types of men. Thus, justice is a matter of knowledge, and not merely the exercise of power in order to fulfil the wishes of the people, since the people may not have knowledge of what is best for themselves.

21 The crisis in Indonesia was part of the larger Asian Financial Crisis which began in July 1997. It began in Thailand when the government, faced with bankruptcy due to its massive foreign debt, switched the national currency from a fixed to a floating exchange rate, causing its collapse. The crisis then spread throughout Asia. In May 1998, Indonesian currency also collapsed, causing enormous inflation and resulting in riots throughout the country and a pogrom against ethnic Chinese, who were blamed for the crisis. Nearly two thousand people were reported to have been killed in the rioting, and there were many rapes of ethnic Chinese women as well.

22 Sir James Michael 'Jimmy' Goldsmith (1933-1997) was a magazine publisher, financier and politician who represented France in the European Parliament between 1994 until his death. He also founded the Referendum Party in the UK. He published a book, *The Trap* (London: Macmillan, 1994), in which he argued that global free trade, which results in widespread competition over cheap labour in the Third World, is a threat to worldwide social stability.

bordering on slavery and rich, insolent and authoritarian bourgeois minorities protected by 'armed forces' from interior repression, thus accelerating environmental destruction, the rise of socio-religious fanaticism, etc. The countries of the South are a powder keg. The recent genocides in central Africa, the increase of bloody civil conflicts in India, Malaysia, Indonesia, Mexico, etc. – which often feed on religious extremism and are sparked by the United States – are only the prelude to a bloody future. Egalitarian ideology masks this reality by congratulating southern countries on their 'democratic progress': deceptive words applied to sham democracies. On the other hand, is it not the case that – by a perverse process (Monnerot's[23] *heterotelia*) – 'democracy' of the Hellene-European sort leads to tragic consequences because of intellectual incompatibilities when it is forcefully imposed upon cultures of the South? To put it briefly, the transplantation of the socio-economic model of the West in these countries proves explosive.

4 – Fourth fracture line, recently described by Jacques Attali:[24] the threat of a **global economic crisis**, a far more serious one than that of the 1930s, and which would be the cause of widespread recession. A forewarning of this may be the fall of the stock markets and currencies in the Far East and the recession now affecting these areas. This economic crisis would have two causes: a) too many countries – and not only poor ones – have more debts than the international banking system can cover (even the debt of European countries has reached worrying levels); and b) the world economy is increasingly based on speculation and the flow of profitable investments (stock markets, trust companies, international pension funds, etc.); this privileging of speculative profits over production carries the risk that a fall of the shares in a given sector could cause a wave of 'widespread panic': international speculators would withdraw their capital, 'drying up' the world economy through a plunging of investments caused by the

---

23 Jules Monnerot (1908-1995) was a French sociologist. He remains largely unknown in the English-speaking world.

24 Jacques Attali (1943-) is a French economist who was an advisor to Mitterrand during the first decade of his presidency. Many of his writings are available in translation. Faye may be referring to Attali's article 'The Crash of Western Civilisation: The Limits of the Market and Democracy', which appeared in the Summer 1997 issue of the American journal *Foreign Policy*. In it, Attali claimed that democracy and the free market are incompatible, writing: 'Unless the West, and particularly its self-appointed leader, the United States, begins to recognise the shortcomings of the market economy and democracy, Western civilisation will gradually disintegrate and eventually self-destruct.' In many ways his arguments resemble Faye's.

A SUBVERSIVE IDEA 63

collapse of the capital market on which companies and states rely. The consequence of this would be a brutal global recession, which would have mournful consequences in a society in which employment is entirely based upon the whims of the economy.

5 – Fifth fracture line: **the surge in fundamentalist religious fanaticism**, particularly in Islam, although Indian polytheists too are contributing to the phenomenon... The rise of radical Islam represents a backlash against the excesses of modern cosmopolitanism, which has sought to impose the idea of atheist individualism, the cult of commodities, the de-spiritualisation of values and the dictatorship of spectacle across the world. As a reaction to this aggression, Islam has taken a more radical form: a dominating and imperialistic role in line with its historical tradition. The number of practicing Muslims continues to grow, while Christianity has lost all traces of militant proselytism and is on the decline – even in South America and Black Africa – after the suicide it committed via the Second Vatican Council,[25] the greatest theological blunder in religious history. Despite reassuring denials on the part of the Western media, radical Islam is spreading like wildfire, threatening new countries: Morocco, Tunisia, Egypt, Turkey, Pakistan, Indonesia, etc. The consequences of this phenomenon will be future civil wars in multi-religious countries like India; violent clashes in Europe – particularly France and Great Britain – where there is a risk of Islam becoming the most practiced religion within twenty years; and an increase in international crises concerning Islamic states, many of which may possesses 'dirty' nuclear weapons. In this respect, it is necessary to denounce the foolishness of those who believe in the idea of a 'Westernised Islam respectful of Republican secularism'. This is unthinkable because Islam is intrinsically theocratic and rejects the very notion of secularism. Conflict, both outside Europe and within it, thus appears inevitable.[26]

6 – A **confrontation between North and South** on theological and ethnic grounds. This is more and more likely to replace the risk that has been avoided so far of a conflict between East and West. No one

---

25 The Second Vatican Council, or Vatican II, was convened in the 1960s in an effort to bring the doctrines of the Church more in tune with the problems of modern life. Many traditionalist Catholics regard it as a surrendering of the Church to secular pressures.

26 It is interesting to note that Faye wrote these words prior to the large-scale Islamist terrorist attacks in the U.S., Spain and the U.K., before the outbreak of the related wars in the Middle East, and also before the mass rioting of Muslims in Paris in 2005, all of which only appear to reinforce his thesis.

can foresee what form this new conflict might take, but it will certainly be a serious one, for it will draw on collective stakes and feelings far stronger than those behind the old antagonism between the United States and the USSR, capitalism and Communism, which was artificial. The current threat is especially fed by the veiled, repressed and dissimulated *resentment* of the countries of the South towards their former colonisers. The *racialisation of the debate* is amazing. Recently, the Prime Minister of an Asian country accused the French government of being 'racist' because it had chosen an Italian investor over one of his country's companies after a simple economic controversy. This racialisation of human relations, the concrete (*heteroclite*) outcome of the 'anti-racist' cosmopolitanism of modernity, is also evident in the West: the Black American leader Farrakhan,[27] like rap musicians across the United States and France (NTM, Ministère AMER, Doc Gynéco, Black Military, etc.), continuously and surreptitiously incite 'vengeance against Whites' and civil disobedience. Paradoxically, egalitarian cosmopolitanism has led to *globalised racism* – an underground and implicit phenomenon that will soon manifest itself openly.

Peoples put beside one another, in close mutual contact in the 'global village' the Earth has now become, are getting ready to clash. And it is Europe, the victim of demographic colonisation, that runs the risk of becoming the main battlefield. Those who envisage the future of humanity as one of widespread race-mixing are wrong: for Europe is the only place where this phenomenon is rife.

The other continents – particularly Africa and Asia – are increasingly forming *impermeable ethnic blocs*, which export their surplus population without importing any from the outside.

A fundamental point: Islam is becoming the symbolic banner of this revolt against the North – a Freudian revenge against 'Western imperialism'. In the collective unconscious of the peoples of the South this leading idea is taking hold: 'mosques are being built on Christian soil' – a revenge for the Crusades and a return to the archaic mindset, history bouncing back like a boomerang. Western and Muslim intellectuals are very wrong when they argue that imperialistic and intolerant fundamentalism is not the very essence of Islam. For the essence of Islam, like Medieval Christianity, is *imperial theocratic totalitarianism*. As for those who offer reassurance by learnedly discussing the present

---

27 Louis Farrakhan (1933-) is the leader of the Nation of Islam, which is the most prominent Black supremacist organisation in the U.S.

A SUBVERSIVE IDEA 65

'lack of unity' among Islamic countries, they should know that these countries will be ready to unite against a common enemy, particularly in emergency situations.

The colonisation of the North by the South presents itself as a form of *soft colonisation*, one that is undeclared and hides behind appeals to solidarity, the right to asylum, and equality. This is the 'strategy of the fox' (as opposed to the lion) which Machiavelli talks about.[28] Actually, the coloniser who justifies his actions by invoking the Western and 'modern' ideology of his victim in no way shares those values he pretends to be adopting. He is anti-egalitarian, domineering (although he presents himself as dominated and persecuted), revanchist and conquering, thanks to the cunning skill of an outlook that has remained archaic. In order to oppose him, must we not *revert to an archaic mindset* ourselves, doing away with the demobilising handicap of 'modern' humanitarianism?

Another cause of the conflict between North and South: *global political and economic competition*. This concerns the war for markets and the control of the scarce resources that are running out (drinking water, fishing resources, etc.), as well as the refusal of the recently industrialised countries of the South to accept anti-pollution measures, and which need to send their surplus population north. History is marked by simple patterns. An insecure, poor and young South with a demographic surplus is pressing against a morally unarmed North that is growing older and older. Let us not forget that the South is arming itself with nuclear weapons, while the North continues to speak of 'disarmament' and 'denuclearisation'.

7 – Seventh catastrophe line: the **unchecked pollution of the planet**, which threatens not the Earth (which still has four billion years to live and can resume the whole process of evolution from scratch) but the physical survival of humanity. This environmental devastation is the result of the liberal and egalitarian (and also formerly Soviet) myth of universal industrial development and an energy-wasting economy for all. Fidel Castro, well inspired for once, in a speech delivered at the WHO[29] of Geneva on 14 May 1998, stated:

'The climate is changing, the oceans and the atmosphere are warmer, the air and waters are contaminated, the soils keep eroding,

---

28 See chapter 1, note 27.
29 World Health Organisation, an agency of the United Nations.

the deserts are expanding, the forests are dying, water is in short supply. Who will save our species? Perhaps the blind and uncontrollable market laws, the neo-liberalisation going global, an economy growing by itself and for itself as a cancer devouring man and destroying nature? That cannot be the way, or it will only be for a very short period of history.'[30] More explicit words could not have been spoken...

What Fidel Castro had in mind when pronouncing these prophetic words was no doubt the irresponsible arrogance with which the United States has refused to reduce its carbon dioxide emissions (at the environmental conferences of Rio de Janeiro and Tokyo).[31] Yet this 'paradoxical Marxist' was also referring to the adhesion of all peoples to an economic model of pure commercial and short-term profit, which is leading everyone to pollute, deforest, destroy the fishing resources of the sea, and plunder fossil fuels and plant resources without any global planning. Fidel Castro was unknowingly appealing here not to Marxism – which is as devastating as capitalism – but to the *ancient Platonic idea of justice.*

8 – It should be added that the 'background' of these seven convergent fracture lines is saturated with aggravating factors that may accelerate the process. Here are a few: the increased vulnerability of techno-economic systems, caused by computer technology (the famous millennium bug of the year 2000);[32] the **proliferation of nuclear weapons** in Asian countries (China, India, Pakistan, Iraq, Iran, Israel, Korea, Japan, etc.) that are in strong conflict with their neighbours and open to nervous and unforeseeable reactions; the weakening of states in the face of the **mafias** that control the traffic of drugs (both natural and – increasingly – synthetic) and are extending it, while also entering new economic sectors, from the war industry to real estate and the agribusiness. According to a recent U.N. report, these global mafias

---

30 The complete text of Castro's address can be found on-line at www.nnc.cubaweb.cu/discur/ingles/14mayo98.htm.

31 The Earth Summit, sponsored by the United Nations, was held in Rio de Janeiro in June 1992. The Kyoto Protocol, which was a further effort by the UN to reduce greenhouse gas emissions, was signed on 11 December 1997 and went into effect in February 2005. As Faye says, the United States, which was responsible for 36.1% of emissions in 1990, has never ratified it.

32 The millennium bug, or Y2K, was a problem that resulted from much of the computer software designed in the late Twentieth century only using the last two digits of the year for dating rather than all four, meaning that at midnight on 1 January 2000 many electronic systems would be unable to tell whether it was 2000 or 1900. Computer software designers went to work on this problem for years prior to the millennium, however, and there were no significant problems when it finally came. However, many experts during the late 1990s were predicting catastrophic consequences for global civilisation after Y2K.

A SUBVERSIVE IDEA

have means beyond those at the disposal of the international forces of repression. Let us not forget the re-emergence of archaic **viral and microbial diseases**, which are eroding the myth of our immunity to epidemics – AIDS was only the first breach. Particularly because of the mutagenic weakening of antibiotics and the massive flux of populations, we are now threatened by the prospect of a worldwide disorder of the health system: recently, in Madagascar, they failed to cure fourteen cases of pulmonary plague.

To put it briefly, should we believe that modernity is about to smash against a wall and that a *global wreck* is inevitable? Maybe not. But maybe... Is the essence of history, its motor, not fuelled by catastrophe? In this case, however, for the first time there is a risk of the catastrophe being a global one, in a globalised world. Already in 1973, the brilliant American ethologist and playwright Robert Ardrey[33] prophesised, 'The modern world is like a train full of ammunition running in the fog on a moonless night with its lights out.'

* * *

These expected catastrophes are the direct consequence of modernity's incorrigible *faith in miracles*: suffice it to consider the myth that a high standard of living could be achieved on a global scale, or the idea of extending economic systems based on high energy consumption to all. The dominant paradigm of materialist egalitarianism – a 'democratic' consumer society of ten billion people in the Twenty-first century *without* any indiscriminate plundering of the environment – is a senseless utopia.

This absurd faith clashes with *physical limits*. Hence, the civilisation it has engendered will not last for much longer. *This is the paradox of egalitarian materialism: it is idealistic and concretely unfeasible.* And this for social reasons – it leads to the dismantling of society – and

---

33 Robert Ardrey (1908-1980) was a widely read and discussed author during the 1960s, particularly his books *African Genesis* (1961) and *The Territorial Imperative* (1966). Ardrey's most controversial hypothesis, known as the 'killer ape theory', posits that what distinguished humans' evolutionary ancestors from other primates was their aggressiveness, which caused them to develop weapons to conquer their environment and also leading to changes in their brains which led to modern humans. In his view, aggressiveness was an inherent part of the human character rather than an aberration. Ardrey's ideas were highly influential at the time, most notably in the 'Dawn of Man' sequence of *2001: A Space Odyssey*, and also in the writings of GRECE, in which Ardrey was frequently cited. They also elicited responses from scholars such as Konrad Lorenz and Erich Fromm. In more recent years, however, Ardrey's theories are no longer upheld by the mainstream scientific establishment.

even more so on the environmental level, given that the planet cannot physically sustain the widespread development of economic forms based on high energy consumption. 'Scientific progress' has missed its mark. This, however, should not lead to a rejection of technology and science, but – as we shall see later on – to their redefinition along inegalitarian lines.

It is not a matter of *whether* the global civilisation built on egalitarian modernity will collapse, but of *when* this will happen. We are thus finding ourselves in an *emergency situation* (what Carl Schmitt referred to as *Ernstfall*, a fundamental concept which he argued liberal egalitarianism never really grasped, as it interprets the world according to a providential and miraculous logic, shaped by the ascending line of progress and development). Modernity and egalitarianism have always refused to believe that an end could come for them as well: they have never acknowledged their own mistakes, pretending to ignore that all civilisations have been – and are – mortal. For the first time certainty exists that the global order of civilisation is threatened with collapse, as it is founded on the spurious and paradoxical idea of *idealist materialism*.

4 – Content: Archeofuturism

Probably only *after* catastrophe will have destroyed modernity, with its global myth and ideology, will an alternative view of the world assert itself by virtue of necessity. No one will have the foresight or courage to implement it before chaos breaks loose.

It is up to us, therefore, who are living in the *interregnum* – to use Giorgio Locchi's expression – to develop the idea of the world for the post-catastrophic age. It may be centred on Archeofuturism, but this concept must be filled with meaning.

1 – The essence of archaism. It is necessary to give the word 'archaic' its true meaning, which is a positive one, as suggested by the Greek noun *archè*, meaning both 'foundation' and 'beginning' – in other words, 'founding impulse'. The word also means 'what creates and is unchangeable' and refers to the central notion of 'order'. 'Archaic' does not mean 'backward-looking', for it is the historical past that has engendered the egalitarian philosophy of modernity that is now falling into ruin, and hence any form of historical regression would be absurd. Modernity already belongs to a past that is over. Is archaism a form of

A SUBVERSIVE IDEA

traditionalism? Yes and no. Traditionalism entails the *transmission of values* and is rightly opposed to those doctrines that wish to make a clean sweep of things. It all depends on what traditions are handed down: universalist and egalitarian traditions are not acceptable, nor are those that are diseased, demobilising and fit only for museums. Should we not draw a distinction when it comes to traditions (values transmitted) between *positive* and *harmful* ones? Our current of thought has always been torn and weakened by an artificial distinction contrasting 'traditionalists' with those 'who look towards the future'. Archeofuturism can reconcile these two families through a dialectic overcoming.

The challenges that shake the world and threaten the downfall of egalitarian modernity are already of an archaic sort: the religious challenge of Islam; the geopolitical and *thalassocratic*[34] battles over scarce agricultural, fishing and energy resources; the conflict between North and South, and colonising immigration into the northern hemisphere; the pollution of the planet and the physical clash between the ideology of development and reality.

All these challenges lead us back to age-old problems. The almost theological political discussions of the Nineteenth and Twentieth centuries, which were like debates concerning the gender of angels, are being cast into oblivion.

This return to 'archaic' (and hence fundamental) questions baffles 'modern' intellectuals, who expound on homosexuals' right to get married and other such inanities. The attraction towards the *insignificant* and the memorialising of the past is a characteristic of dying modernity. *Modernity is backward-looking, whereas archaism is futurist.*

On the other hand, as foretold by philosopher Raymond Ruyer – someone hated by Leftist intellectuals – in his seminal works *Les nuisances idéologiques*[35] and *Les cents prochains siècles*,[36] when the historical period of the Nineteenth and Twentieth centuries will have come to

---

34 A thalassocracy is a state which depends primarily on the sea for its power, either economically or strategically. The Greek historian Herodotus described ancient Phoenicia as a thalassocracy, since it controlled little territory on land but possessed a large network of city-states which flourished through maritime trading.

35 *Les nuisances idéologiques* (Paris: Calmann-Lévy, 1971), or *The Ideological Nuisances*. It has never been translated.

36 *Les cents prochains siècles* (Paris: Fayard, 1976), or *The Next Hundred Centuries*. It has never been translated.

a close, and its egalitarian hallucinations will have been sunk by catastrophe, humanity will revert to its archaic values, which are purely biological and human (i.e., anthropological): the separation of gender roles; the transmission of ethnic and folk traditions, spirituality and priestly organisation; visible and structuring social hierarchies; the worship of ancestors; rites and tests of initiation; the re-establishment of organic communities (from the family to the folk); the de-individualisation of marriage (unions must be the concern of the whole community and not merely of the married couple); an end of the confusion between eroticism and conjugality; the prestige of the warrior caste; inequality among social statuses – not implicit inequality, which is unjust and frustrating and is what we find today in egalitarian utopias, but explicit and ideologically legitimated inequality; duties that match rights, hence a rigorous justice that gives people a sense of responsibility; a definition of peoples – and of all established groups or bodies – as diachronic communities of destiny rather than synchronic masses of individual atoms.

In brief, in the vast, oscillating movement of history which Nietzsche called 'the eternal return of the identical',[37] future centuries will witness a return to these archaic values one way or another.

The problem for us Europeans is not having these values imposed upon us, on account of our cowardliness, by Islam – as is already happening – but rather of being capable of asserting these values ourselves by drawing them from our historical memory.

Recently, a great patron of the French press, whose name I cannot mention, and who is known for his social-liberal views, shared the following disenchanted thought with me: 'In the long run, the values of the market economy will lose against those of Islam, for they are exclusively based on individual economic profit, and this is inhuman and transient.' It is up to us to make sure that it won't be Islam that will impose an inevitable return to reality upon us.

It is evident that the ideology in power today – and not for much longer – considers these above-mentioned values *diabolical*, just as a paranoid madman might see the psychiatrist that is curing him as the

---

37 'What if some day or night a demon were to steal into your loneliest loneliness and say to you: "This life as you now live it and have lived it you will have to live once again and innumerable times again; and there will be nothing new in it, but every pain and every joy and every thought and sigh and everything unspeakably small or great in your life must return to you, all in the same succession and sequence..."' From Friedrich Nietzsche, *The Gay Science* (Cambridge: Cambridge University Press, 2001), p. 194. This is one of Nietzsche's central ideas.

A SUBVERSIVE IDEA

devil. Actually, these are the *values of justice*. Forever suited to human nature, these values *reject the erroneous idea of individual emancipation* promoted by the philosophy of the Enlightenment, which leads to the isolation of man and social barbarism. These archaic values are *just* in the ancient Greek sense of the term, for they see man for what he is, a *zoon politikòn* ('social and organic animal within a communitarian city') rather than for what he is not – an asexual and isolated atom possessing universal and enduring pseudo-rights.

Concretely, these anti-individualist values enable the attainment of self-realisation, active solidarity and social peace, whereas the falsely emancipating individualism of egalitarian doctrines brings the law of the jungle.

2 – The essence of futurism. A constant feature of the European mindset is the rejection of what is unchangeable: a Faustian, (at)tempting character (in the sense of one who both 'makes attempts' and 'makes one undergo temptations'), which embarks upon new forms of civilisation. The European cultural background America has inherited is adventurous and – most importantly – *voluntaristic*. It aims to change the world through the creation of empires or technological science, by means of *vast plans* that represent the *anticipated representation of a constructed future*. The 'future', as opposed to a historical cycle that repeats itself, is what lies at the centre of the European worldview. To paraphrase Heidegger, it could be said that history is like a path that unwinds through a forest (*Holzweg*),[38] or rather the course of a river along which one must always face new dangers and make new discoveries. Besides, according to this futurist view, technological and scientific inventions, just like political or geopolitical projects – regarded as challenges – are approached from an aesthetic as well as utilitarian angle. Aviation, rockets, submarines and nuclear power have sprung from *rationalised fantasies* where the scientific spirit has managed to carry out the plan conceived by the aesthetic.

The European soul is marked by a longing for the future, a sign of *youthfulness*. To put it shortly, it is *historial and imaginal* (it constantly envisages future history according to a plan).

---

38 '"Wood" is an old name for forest. In the wood there are paths, mostly overgrown, that come to an abrupt stop where the wood is untrodden. They are called *Holzwege*. Each goes its separate way, though within the same forest. It often appears as if one is identical to another. But it only appears so. Woodcutters and forest keepers know these paths. They know what it means to be on a *Holzweg*.' From Martin Heidegger, *Off the Beaten Track* (Cambridge: Cambridge University Press, 2002), p. v.

In art, too, European civilisation has been the only one in which forms have undergone constant renovation and all cyclical return of past models has been banned. The spirit of artworks must remain unchanged (the archaic pole) but their form must always change (the futurist pole). The European soul is defined by ongoing *creation* and *invention* – the *poiesis*[39] of the Greeks – while being always aware of the fact that in its direction and values it must remain faithful to tradition.

The essence of futurism is the *planning of the future* (not 'making a clean sweep of the past'); the envisaging of civilisation – in this case, European civilisation – as a *work in motion*, to paraphrase Wagner's[40] musical expression. Politics here are understood not merely in a narrow sense as the 'identification of one's enemy' (Carl Schmitt), but as the *identification of one's friend (*who is part of the folk community?) and – most importantly – as the future *transformation of the folk*, driven by ambition, a spirit of independence, creativity and the will to power...

This dynamic force, however, and projection towards the future, meets many obstacles. The first is egalitarian modernity with its morality – which lays guilt upon force – and its historical fatalism. The second obstacle, or rather danger, in the social field is represented by a deviated form of futurism which may lead to utopian aberrations for the sheer taste of 'change for the sake of change'. Thirdly, when left to itself – particularly in the realm of technological science – the futurist mentality may prove suicidal, especially because of its impact on the environment, given the risk of deifying technology as something that can 'solve everything'.

Hence, futurism must be *tempered* with archaism; or, to use a bold expression, we might say that *archaism must cleanse futurism*.

The futurist mindset has also encountered a number of 'barriers': a limit to space-based enterprises because of their high cost, the trivialising of technological science and its loss of meaning, disenchantment towards all positive and 'creative' values of mobilisation, widespread loss of poetic and aesthetic qualities through commercialisation, etc.

---

39 Ancient Greek: 'to make'. It is the etymological root of the word poetry. Plato, in his 'Symposium', defined *poiesis* as the method by which mortals attempt to transcend death, such as through sex, fame or knowledge.

40 Richard Wagner (1813-1883) was the greatest German composer of operas in the Nineteenth century (although he preferred to call his mature works 'music dramas'). The influence of his music and writings has had a tremendous influence on all aspects of culture in the West.

A SUBVERSIVE IDEA

The implication of all this is that futurism can only become a driving force if it takes a *new course*. The neo-archaic world that is looming near is the only one capable of freeing the futurist spirit from the impasses of modernity.

3 – **The Archeofuturist synthesis** as a philosophical alliance between the Apollonian and the Dionysian.[41] Futurism and archaism are both related to Apollonian and Dionysian principles that have always appeared to be mutually opposed, when in fact they are complementary. The futurist pole is Apollonian in its sovereign and rational plan to shape the world, and Dionysian in its aesthetic and romantic mobilisation of pure energy. Archaism is telluric in it appeal to timeless forces and conformity to the *archè*, but it is also Apollonian, for it is founded on wisdom and the endurance of human order.

It is a question, for future society, of no longer thinking according to the exclusive logic of '*or*' but according to the inclusive logic of '*and*'; of simultaneously embracing *ultra-science* and a return to traditional solutions that date back into the mists of time. Futurism is actually more vigorous than archaism: for reasons of sheer realism, *a futurist plan can only be implemented by resorting to archaism*.

Hence the paradox of Archeofuturism, which rejects all ideas of progress, as everything pertaining to the worldview of a people must rest on unchangeable bases (although forms and expressions may vary): for over the past 50,000 years *homo sapiens* has changed very little, and archaic and pre-modern models of social organisation have proven valid. The fallacious idea of progress must be replaced with *movement*.

An astonishing degree of continuity exists between archaic values and the revolutions technological science makes possible. Why? Because the egalitarian and humanitarian mindset of modern man, for instance, does not allow him to manage the explosive possibilities behind genetic engineering or the new electromagnetic weapons (in the making). The incompatibility between modern egalitarian ideology and futurism emerges in the extraordinary limits placed upon the civil nuclear power industry in the West through the influence of manipulated public opinion, or in the pseudo-ethical obstacles raised

---

41 The Apollonian/Dionysian dichotomy was first coined by Nietzsche in his early work, *The Birth of Tragedy*. He defined the Apollonian as that which was related to dreams, rational, and most apparent in the visual arts. He understood the Dionysian as intoxication, being passionately tied to the instincts, and best seen in music.

in opposition to genetic engineering, the creation of 'modified' human beings, and positive eugenics.

The more archaic futurism becomes, the more radical it will be; the more futurist archaism becomes, the more radical it will be.

Needless to say, Archeofuturism is based on the Nietzschean idea of *Umwertung*[42] – the radical overthrowing of modern values – and on a spherical view of history.

Egalitarian modernity, founded as it is on faith in progress and boundless development, has adopted a secular version of the *linear, ascendant, eschatological and soteriological* (redemptive) view of history, which stretches back to the time of the religions of salvation, and which is also shared by socialist and liberal democratic thought. Traditional societies (particularly non-European ones) have developed a cyclical, repetitive and hence fatalistic view of history. The Nietzschean view of history, which Locchi described as 'spherical', differs from both the linear and the cyclical notions of progress.

So what is this view?

Let us imagine a sphere, a billiard ball moving in disorderly fashion across a surface, or moved by the (necessarily imperfect) will of a player: after a number of spins, the same point on the surface of the ball will inevitably touch the cloth. This is the 'eternal return of the *identical*', but not of the 'same'. For the sphere is moving and even if that very 'same' point is touching the cloth, its position is not the same as before. This represents the return of a 'comparable' situation, but in a different place. The same image can be applied to the succession of the seasons and the historical outlook of Archeofuturism: the return to archaic values should not be understood as a cyclical return to the past (a past that has failed, as it has engendered the catastrophe of modernity), but rather as the *re-emergence of archaic social configurations in a new context*. In other terms, this means applying age-old solutions to completely new problems; it means the reappearance of a forgotten and transfigured order in a different historical context.

Three additional points of a philosophical nature are in order. The first: Archeofuturism distinguishes itself from conventional

---

42 *Umwertung aller Werte*, or 'transvaluation/revaluation of all values'. This was a key concept in Nietzsche's last works. He wrote: 'Let us not underestimate the fact that *we ourselves*, we free spirits, already constitute a "revaluation of all values", a living declaration of war on and victory over all old concepts of "true" and "untrue".' From *The Anti-Christ, Ecce Homo, Twilight of the Idols and Other Writings* (Cambridge: Cambridge University Press, 2005), p. 11.

A SUBVERSIVE IDEA

'traditionalism' because of its approach to technological science, which it does not demonise: for the essence of technological science is not connected to egalitarian modernity, but rather has its roots in the ethno-cultural heritage of Europe, and particularly ancient Greece. Let us remember that the French Revolution 'did not need any scientists',[43] so much so that it guillotined several of them.

Second point: Archeofuturism is a *changing worldview*. The values of the *arché*, projected into the future, are made newly relevant and transfigured. The future is not the negation of the tradition and historical memory of a folk, but rather their *metamorphosis*, by which they are ultimately reinforced and regenerated. To use a metaphor: what does a nuclear-powered ballistic missile submarine have in common with an Athenian trireme? Nothing and everything: one represents the metamorphosis of the other, but both, in different ages, have served precisely the same purpose and embody the same values (including the same aesthetic values).

Third point: Archeofuturism is a *concept of order*, a concept that upsets modern minds, which are shaped by the fallacious individualist ethics of emancipation and the rejection of discipline that has led to the swindle of 'contemporary art' and wreaked havoc in the educational and socio-economic systems.

According to the view of Plato that he conveyed in *The Republic*, order is not injustice: *Every conception of order is revolutionary and every revolution is a return to authentic order.*

4 – The concrete applications of Archeofuturism. A concept for which no *examples* can be given is not an effective one. Marxism has partly failed because Marx and Engels, caught up in the 'philosophy of no' and ultra-criticism, failed to provide a concrete description – however brief – of their 'Communist society'. The result: while its critique of capitalism was often a pertinent one, the Communist paradigm has been implemented in an improvised manner, often under the leadership of autocrats and tyrants. Communism has collapsed because, despite being an ideology radically opposed to the bourgeois order, it remained an *abstract logic of resentment* that people have attempted to put into practice through hastily drawn political dogmas. Today, new paths must be paved:

---

43 This quotation is attributed to various officials of the Revolutionary Tribunal during the French Revolution, which sent many people to the guillotine. The occasion was the sentencing of the chemist Antoine Lavoisier, often called the 'father of modern chemistry', to death in 1794.

A. An answer to the approaching confrontation between North and South and the rise of Islam. This global return to the archaic that began in the 1980s has radically altered modern geopolitics: Islam has once more embarked on its march of conquest, which European colonisation had interrupted a few centuries ago; colonising migrations are pouring into the northern hemisphere like a backlash against colonisation and the demographic ageing of the North; the Nineteenth and Twentieth century opposition between Europe and North America, and – within the Eurasian continent – between 'Westerners' (which did not always include Germans) and Slavs is coming to an end. Today's contrast – tomorrow's confrontation – is between North and South. We are already facing Archeofuturist challenges.

Yielding to the naive myth of 'interracial integration' or ethnopluralist 'communitarianism' is an aberration.

The mindset of Muslims and immigrants from the South, as well as that of the sons of the immigrants who, in expanding and increasingly aggressive masses, are inhabiting European cities, as well as that of the leaders of the emerging Muslim and Far Eastern powers, while masked by a hypocritical Western and modern gloss, has remained archaic: it is based on the primacy of force, the legitimacy of conquest, exacerbated ethnic exclusivity, aggressive religiosity, tribalism, machismo, and a worship of leaders and hierarchic order – although it is disguised as democratic Republicanism.

We are witnessing the return of wide-scale invasions under a new guise. The phenomenon is far more serious today, as the 'invaders' have preserved a formidable 'home base': the countries they have left, the motherlands which are always solidly behind them and ready to defend them – and which secretly aspire to do so through force in the future. This is why I am speaking in terms of colonisation rather than invasion.

The modern egalitarian mindset is utterly powerless. Would it not be better, then, to readopt those archaic values that inspire our very real enemies and which – significant differences notwithstanding – have remained the same for all peoples, before and after the interlude of modernity?

B. The answer to the decline of nation-states and the challenge of European unification. In this respect, it is essential to prepare for a likely confrontation by doing away with the modern altruism of universal harmony.

A SUBVERSIVE IDEA

It is a matter of *rethinking war*, not in its modern form as war between nations, but as it existed in Antiquity and the Middle Ages: as the clash between vast ethnic or ethno-religious blocs. It would be interesting to reconsider – in the new forms in the making – the kind of macro-solidarity once embodied by the Roman Empire and European Christendom, and to pragmatically define the idea of *Eurosiberia* as a block extending from Brest to the Bering Strait, from the Atlantic Ocean to the Pacific, across fourteen time zones: a land where the sun never sets and thus the largest geopolitical unit on Earth. Russian leaders are already thinking about this[44] – in uncertain terms and through the fumes of vodka, but still: they are thinking about it. It would be worth asking ourselves whether French nationalism may not be completely outdated, whether the nation-state in Europe may not be as anachronistic as Maurras'[45] monarchist movement was in the 1920s, and whether the groping and tentative construction of a federal European state – for all its short-term inconveniences – in the long run may not prove the only means, as a revised adaptation of the Roman and Germanic imperial model, of preserving the brother-peoples of our Great Continent from oblivion.

It is also worth asking ourselves whether in this context the United States still represents an *enemy* (as I myself once argued) – which is to say, a power posing a mortal threat – rather than a *foe* or economic, political and cultural *rival*. To raise this question is to identify the neo-archaic problem of the global solidarity of the North – which is essentially ethnic in nature – against the threat of the South. In any case, *the notion of the West is disappearing and being replaced by the idea of the Northern World or the North.*

As in the Middle Ages or Antiquity, the future requires us to envisage the Earth as structured in vast, quasi-imperial units in mutual conflict or cooperation.

---

44 Faye is probably referring to the Eurasianist Movement in Russia, a concept which dates back to White Russian émigrés of the 1920s and which was most notably revived by the political philosopher Aleksandr Dugin in the 1990s. It is a geopolitical theory, often seen as a corollary of Dugin's ideology of National Bolshevism, which asserts that Moscow, Berlin and Paris form a natural geopolitical axis that Dugin believes must be realized in order to bring about a revolt against American world domination. Dugin's Eurasia Party was officially recognized by the Russian government in 2001, and it was widely rumoured to have sympathizers at the highest levels of Vladimir Putin's administration, although in subsequent years Dugin has been critical of Putin.

45 Charles Maurras (1868-1952) was a French Catholic counter-revolutionary philosopher who was the founder of the *Action Française* (see chapter 1, note 12).

Is it not the case that the future belongs to a *neo-federal Europe founded on autonomous regions*, a contemporary version of the ancient and Medieval organisation of the continent? And this for the simple reason that an enlarged techno-bureaucratic Europe comprised of twenty-odd uncertain and divided nations of substantially different sizes would merely be an apolitical jumble under the control of the United States and NATO, one open to immigrant colonisation and uncontrolled competition from the new industrial countries. After the Euro – the first return to a continental currency since the end of the ancient world – can we now envisage the *United States of Europe*, a vast federal power open to an alliance with Russia?

C. **The answer to the crisis of democracy.** Peter Mandelson,[46] the person behind Tony Blair's New Labour in Britain, and Wolfgang Schäuble,[47] Kohl's[48] Christian Democrat rival, held a series of meetings in March 1998 to discuss the 'future of democracy' which were reported in the London newspaper *The Guardian*. Schäuble was bewildered and did not always agree with the brilliant and 'Leftist' British political theorist.

Here is a quote from Mandelson: 'It may be that the era of pure representative democracy is coming slowly to an end. ... Democracy and legitimacy need constant renewal. They need to be redefined with each generation. ... Representative government is being complemented by more direct forms of involvement from the Internet to referendums. This requires a different style of politics and we are trying to respond... People have no time for a style of government that talks down to them or takes them for granted.'[49]

Schäuble, struck by such populist and 'antidemocratic' daring, makes the following comment: 'I think that we politicians have to take the decisions. In short, Mr. Mandelson's verdict is: "Representative democracy is over". Translated, that means, "Things must be brought

---

46 Peter Mandelson (1953-) was the M.P. for Hartlepool from 1992 until 2004, and helped to rebrand the Labour Party as 'New Labour', which was key to Blair's electoral victory in 1997. He served in Blair's cabinet.

47 Wolfgang Schäuble (1942-) was a member of Kohl's cabinet between 1984 and 1991, and chairman of the Christian Democrat group in parliament between 1991 and 2000. He is currently Federal Minister of the Interior. He was very popular in Germany during the 1990s and was widely speculated to be Kohl's successor as Chancellor, but as the Christian Democrats were defeated in the 1998 election this never came to pass.

48 Helmut Kohl (1930-) was Chancellor of West Germany between 1982 and 1990, and then became the first Chancellor of reunited Germany between 1990 and 1998.

49 This was said at a seminar held at the British Embassy in Bonn on 15 March 1998.

closer to the people". That means politicians are too cowardly to take decisions. Mandelson also argued that if Europe is to function at all, then it can only be through inter-governmental co-operation. That's the end of European integration if you don't want to lead politically and take decisions.'

It would be hard to imagine a more pointed attack on the 'modern' model of Western parliamentary democracy, which was theorised by Rousseau in his *Social Contract* and has now grown obsolete. Anglo-Saxon pragmatism often makes ideological openings – however ill-defined – possible which are completely ruled out by French doctrinalism, German Idealism and Italian Byzantinism.

Mr. Mandelson, a distinguished New Labour egghead, is an Archeofuturist without knowing it. For he is telling us that the 'modern' parliamentary democracy we inherited from Eighteenth and Nineteenth century paradigms will be unsuited to the world of the future. Slow and weak decision-making, compromises and the lack of an authority capable of asserting itself in 'emergency cases' are increasingly common features, as are the dictatorship of bureaucracies and speculators, the paralysis of parliaments, the corrupt career-making of party members, the growth of mafias, etc.

Modern democracy defends not the interests of the people, but those of illegitimate minorities. It distrusts the people and discredits the idea of 'populism' by equating it with dictatorship, which is really absurd. Without any ideological or pseudo-moral prejudices, Mandelson also suggests the need to restore a daring and decisive form of political authority, yet one resting on the will of the people, particularly thanks to 'more direct forms of involvement from the Internet to referendums.'

These suggested paths are all very interesting, for they seek to reform democracy by combining two archaic elements with a futurist one.

The first archaic element: the sovereign decision-making power set in motion by the *direct* will of the people. This brings to mind the model of *auctoritas*[50] of the first Roman Republic, as symbolised by the initials SPQR (*Senatus Populusque Romanus*, 'The Senate and People of Rome'):[51] a close link between popular aspirations and established

---

50 Latin: 'authority'.
51 SPQR was meant to embody the idea that the government of the Roman Republic represented the rule of the people. It continued to be used during the Roman Empire and by Fascist Italy, and remains the motto of the city of Rome to this day.

authority, which imposes its decrees without being censored by any judges or 'law' above the will of the people. It would also be possible to refer to the model of Fourth or Fifth century BC Athens or the structure of Germanic tribes.

Second archaic element: the reconciliation of political institutions with the general population. The modern nation-state, as originally conceptualised by Hobbes, has distanced the people from its sovereignty through the illusion of a better representation of popular will. Labour MP Mandelson is implicitly suggesting a return to Athenian, Roman and Medieval principles, through a closer link between the people and its leaders. On the other hand, the term *demos* ('democracy': 'power of the *démoi*') literally means 'neighbourhood' or 'rural district'. In this respect, it would be possible to imagine a decentralised Europe where 'local peoples' establish their own laws, according to the Imperial Roman or Medieval Germanic model.

Third element, a futurist one: the possibility of directly voting at referendums by e-mail or using individual encrypted codes. Fearing the common masses, the political and media establishment rejects this solution, afraid that its manoeuvres may be exposed. Here too – as in the field of biology – the dominating ideology of modernity is fighting and censuring in order to limit the possibilities offered by technological science. *Modernity is reactionary*.

But what is a people and what will it be in the future?

Is a people *laios*, the 'mass' dear to Marxists and liberals, i.e., the 'present population', based on the law of territory, or is it *ethnos*, a folk community founded on the law of blood, culture and memory? Modernity tends to define a people as *laios*, a rootless mass of individuals coming from all different places. But the future which is inexorably looming near is reawakening ethnic loyalty and tribalism both on a local and global scale. Tomorrow a people will return to be what it has always been, prior to the short interlude of modernity: *ethnos*, a community both cultural and biological. I insist on the importance of biological kinship to define peoples, and particularly the family of European peoples (as well as all others), not only because humanity – contrary to what the melting-pot myth suggests – is increasingly defining itself through 'ethno-biological blocs', but also because the inherited characteristics of a people shape its culture and outlook.

D. The answer to social disintegration. That collapse is looming close can be seen from the failure of educational systems, which are no

A SUBVERSIVE IDEA

longer able to curb illiteracy and crime in schools, for they are dominated by the illusion of 'non-authoritarian' methods of teaching; this can be seen in the spread of urban crime, which is caused not only by unrestrained immigration, but also by the unrealistic dogma of deterring crime through education and by the obliteration of the ancient *principle of repression* – something far from tyrannical when it is based on law. It can also be seen from the demographic collapse caused both by anti-natalist governmental policies and by the *ethnic masochism* of the ruling ideology, as well as by the exacerbated hedonistic individualism that is triggering a boom in anti-natural practices: divorces made automatic – and which will soon be mere administrative formalities – the ridiculing and obstinate rejection, both fiscal and social, of the housewife model, the spread of short-lived and sterile forms of common law marriages, the glorification of homosexuality and soon of legal gay marriages (that will enable those in such unions to be able to adopt children), etc. The demographic fall caused by anti-natalism will lead to economic disaster in Europe by 2010, because of the growing deficit in social budgets caused by the ageing of the population.

Everywhere modernity, which appears to be triumphing, is actually already languishing and failing in its attempt at social regulation: for, as Arnold Gehlen[52] grasped, it is based on a dream-like view of human nature and fallacious anthropology.

It is likely that the *post-catastrophic world* will have to reorganise social fabrics according to archaic principles – which is to say, human ones.

What are these principles? The power of family units, which are invested with authority and have responsibilities towards their offspring; the legal primacy of the *principle of punishment* over prevention; the subordination of rights to duties; the framing – not recruitment – of individuals within communitarian structures; the power of social hierarchy, newly made visible through solemn social rites (aesthetic-magical function); the rehabilitation of the *aristocratic principle*, which is to say of the rewards given to the best and most worthy (for courage, service and skill), in the awareness that a surplus of rights

---

52 Arnold Gehlen (1904-1976) was a German philosopher who was active in the Conservative Revolution. He joined the Nazi Party in 1933 and remained in its ranks until the end of the war, being drafted into the Wehrmacht in 1943. After denazification, he continued to write and teach after the war, and his ideas remain influential on the German Right to this day. His post-war books *Man in the Age of Technology* and *Man, His Nature and Place in the World* have been published in English.

corresponds to a surplus of duties and that aristocracies should never degenerate into plutocracies and be wary of becoming hereditary.

Is it then a matter of 'abolishing freedom'? Paradoxically, it is 'emancipating' modernity that has destroyed concrete freedoms by proclaiming an abstract Freedom. While in Europe it is nearly impossible to expel illegal immigrants, the mafias are branching out, criminal gangs enjoy ever greater impunity, and citizens who respect the social pact are increasingly being recorded in police records, monitored and having their finances checked, sanctioned and bled by tax authorities.

Faced with this failure, would it not be better to restore concrete Medieval or ancient institutions such as *franchises, local communitarian pacts and forms of organic solidarity among neighbours*?

These, then, are the general principles. They will probably serve as the foundations for the future societies that will emerge from the rubble of modernity. It is up to the *new ideologues* of our current of thought to define these principles and concretely implement them. A few concrete questions should already be raised.

In a random order: Why keep schooling compulsory until the age of sixteen rather than limit ourselves simply to primary school, where through discipline basic subjects could effectively be taught? Kids over thirteen would then be free to choose whether they wish to start working as apprentices or continue their studies. In such a way we would overcome the impasse of the current system, which leads to failure, uncivil behaviour, ignorance, semi-literacy and unemployment. A well-organised and rigorous primary-school system would undoubtedly produce young people of a higher level than the often quasi-illiterate individuals who are now making it through the collapsing high-school system. All discipline brings freedom. Why should a two-tier educational system, based on severe selection and the assignment of grants – which would prevent plutocracy and the dictatorship of money – be wrong, if it leads to the turnover of elites and to meritocracy?

The new societies of the future will finally abolish the aberrant egalitarian mechanism we have now, whereby 'everyone aspires to become an officer', or a cadre or diplomat, even though all evidence suggests that most people do not have the skills to fulfil these roles. This model engenders widespread frustration, failure and resentment. The societies that will be vivified by increasingly sophisticated technologies, in contrast, will ask for a return to the archaic inegalitarian and hierarchical norms, whereby a competent and meritocratic minority is rigorously selected to take on leading assignments. Those who will perform 'subordinate' functions in these inegalitarian societies will not feel frustrated: their dignity will not be called into question, for *they will accept their own condition* as something useful within the organic community – finally freed from the *individualistic hubris* of modernity, which implicitly and deceptively states that each person can become a scientist or prince.

Another example concerns the treatment of those who commit crimes. The future will force us to rethink the modern and ineffective means of crime prevention and the reintegration of criminals into society by implementing a juridical revolution to restore the archaic methods of repression and forced re-education. Here too we must change the way we think.

To sum up, with the introduction of 'hypertechnologies' the social models of the future will lead us not towards greater egalitarianism (as the stupid apologists of universal communication believe will happen thanks to the Internet), but rather to a return to archaic and hierarchical social models. On the other hand, it is global technological competitiveness and the economic war for the control of markets and scarce resources that are pushing us in this direction: those who will win will be the peoples with the strongest and best selected 'elite blocs' and the most organically integrated masses.

E. The answer to the (global) incapacity for making decisions, the inadequacy of the U.N. 'machine' and the risk of widespread confrontations. The nation-states of the U.N. – from the United States to the Fiji Islands – are incapable of managing the overcrowded spaceship the Earth has become. This was clearly seen at the Tokyo Summit, where these states failed to reach any common agreement to avoid the environmental catastrophes looming near.

As a medium-term solution, it would thus be necessary to organise the planet according to a few, vast 'neo-imperial' units capable of reaching decisions and negotiating with one another. This would mean a return to the ancient world order, which was based on such blocs, albeit under a different form.

The scenario would be as follows: a Sino-Confucian block, a Euro-Siberian unit, an Arab-Muslim one, a North American one, a Black African one, a South American one and finally one including the Pacific and peninsular Asia.

F. The answer to economic and environmental chaos. As we have seen, the modern economic paradigm based on the belief in miracles will meet insurmountable physical obstacles. The utopia of 'development' open to ten billion people is environmentally unsustainable.

The foreseeable collapse of the global economy allows us to envisage and formulate the hypothesis of a revolutionary model based on a self-centred and inegalitarian world economy, which may be imposed upon us by historical events, but which it would be wise to foresee and plan for in advance. This hypothesis is based on three great paradigms. Here is the Archeofuturist scenario:

First off, most of humanity would revert to a pre-technological subsistence economy based on agriculture and the crafts, with a neo-medieval demographic structure. The African population, like that of all other poor countries, would be fully involved in this revolution. Communitarian and tribal life would reassert its rights. 'Social happiness' would most probably be greater than it is in jungle-countries like Nigeria or mega-slums like Kolkata and Mexico City today. Even in industrialised countries – India, Russia, Brazil, China, Indonesia, Argentina, etc. – a significant portion of the population could return to live according to this archaic socio-economic model.

Secondly, a minority percentage of humanity would continue to live according to the techno-scientific economic model based on ongoing innovation by establishing a 'global exchange network' of about a billion people. A considerable advantage of this would be a strong reduction in pollution. Besides, it is difficult to envisage any other solution that would ensure the salvation of the global ecosystem, for even in the near future it will be impossible to make any wide-scale use of clean energy sources.

Finally, these vast, neo-archaic economic blocs would be centred upon a continental or multi-continental plan, with basically no mutual exchange between them. Only the techno-scientific portion of humanity would have access to the global exchange.

This two-tier world economy thus combines archaism and futurism. The techno-scientific portion of humanity would have no right to intervene in the affairs of the neo-medieval communities that form the majority of the population, nor – most importantly – would it in any way be obliged to 'help' them. No doubt, this presents a monstrous picture to the modern and egalitarian spirit, yet in terms of actual collective well-being – which is to say justice – a revolutionary scenario of this sort may prove rather pertinent.

A SUBVERSIVE IDEA

On the other hand, freed from the economic burden of areas 'to be developed' and 'helped', the minority portion of humanity would live in a techno-scientific economic system where innovation would take place at a far higher speed than it does now. Here too, *the return to archaism can be seen to foster futurism and vice versa.*

This is only a sketch, an outline. It would be the job of economists to carry on this reflection.

G. The revolution in biotechnologies. It is in the sphere of biotechnology that the need for Archeofuturism appears most evident. Modern and egalitarian ways of thinking, caught up as they are in the guilt-engendering pitfall of human-rights 'ethics', are incapable of dealing with biotechnological progress and face moral obstacles that are actually para-religious in nature. In such a way, *modernism ends up being anti-scientific.* It hinders the development of genetic and transgenetic engineering. Paradoxically, only neo-archaic ways of thinking enable the use of the genetic technologies that are constantly being curbed today. The modern outlook runs against a substantial obstacle: anthropocentrism and the egalitarian sacralisation of human life, which it inherited from secularised Christianity.

Let us consider the numerous ways in which the biotechnology that is already being developed could be used, now that the stage of animal experimentation is over.

Technologies related to positive eugenics would make it possible not only to cure genetic diseases, but also to improve – by transgenic means – the hereditary performance of individuals, according to chosen criteria. We should also mention the (now imminent) application to man of a process that has already been successfully tested upon animals: the creation of inter-specific hybrids, 'human chimeras' or 'para-humans' that would find countless applications.

Two American researchers have already patented such practices,[53] which have been blocked by politically correct 'ethical committees'. Man-animal hybrids or semi-artificial living creatures would have countless uses, as would decerebrated human clones, which could be

---

[53] A patent application for a technique to produce a human/chimpanzee hybrid was filed by Stuart Newman, a Professor of cell biology, and Jeremy Rifkin, a biotechnology activist, in 1997. However, after a lengthy debate, the U.S. Patent Office rejected the patent in 2005 on the grounds that the 13th Amendment (the abolition of slavery) prohibits the patenting of humans. Prof. Newman said he was actually overjoyed by this defeat, since he had never intended to produce the hybrids, but had used the application, in anticipation of its rejection, as a means to establish a legal precedent to prevent patents being issued on living things, as the Patent Office had already issued patents to several other products of genetic engineering.

used as organ banks. This would put a stop to the odious traffic that particularly affects the poor people in Andean America.

Let us also recall the potential human application of a technology that has already been tested on sheep in Scotland: 'birth without pregnancy' through the development of embryos in an artificial amniotic environment ('incubators').

Clearly, those who support modern ideology consider the mere mention of such technologies as something Satanic. Yet, their use is becoming possible... Is it better then to brutally censure a scientific discovery or to carefully consider its social applications?

H. The Archeofuturist ethic. Archeofuturism would enable us to do away with the scourge of egalitarian modernism, which is hardly compatible with the century of iron that awaits us: the *weak spirit* of humanitarianism, a sham ethic which raises 'human dignity' to the rank of a ridiculous dogma. This, not to mention the hypocrisy of the many well-meaning souls who yesterday forgot to denounce Communist crimes and today have nothing to say about the embargo on Iraq and Cuba by the American superpower,[54] Indian nuclear tests, the oppression of the Palestinians, etc.

This spirit serves as a means of moral disarmament, for it establishes paralysing proscriptions, taboos that engender guilt and concretely prevent European public opinion and leaders from facing the present threats.

Actually, what is promoted and implemented under the guise of moral principles is a Leftist policy that aims to destroy the very European substratum of Europe. For instance, the campaign against the (legal) deportation of *'sans-papiers'* – which is to say illegal immigrants – led by the French intelligentsia and show business' efforts to make the deportation of any immigrant impossible in the name of human rights and the pseudo-principles of charity. The underlying ideology and true strategic objective here is – according to a neo-Trotskyist plan – the flooding of Europe with the surplus population of peoples from the South.

A further dilemma: the campaigns against the nuclear power industry, which are leading to the dismantling of Swedish and German plants and to the complete abandonment of nuclear power by some of

---

54 The United Nations imposed sanctions on Iraq, following its invasion of Kuwait, on 6 August 1990. Between 1990 and 2003, when the sanctions were lifted following the U.S. invasion, the sanctions caused great misery for the Iraqi population. There was a huge increase in child mortality, with estimates of the number of children who died as a result of the sanctions running in the hundreds of thousands.

A SUBVERSIVE IDEA

the European states, with the exception of France, which continues to resist – but for how much longer? Everybody knows that, controllable accidents notwithstanding, nuclear energy is the least polluting among the energies currently available.

This operation, too, aims at *weakening Europe* through the excuse of humanitarianism, by depriving it of the leading energy technologies, economic independence, and – at the same time –of any integrated form of nuclear deterrence. The stimulus behind this manipulation, of which the credulous intellectual and artistic bourgeoisie of Europe has been made a victim, is a sort of monstrous and irresponsible exaggeration of the maxim 'love thy neighbour like thyself' – an apology for weakness and a pathological form of emasculation and self-blame. What we are facing here is a sub-culture of emotionality, a cult of decline that serves to lobotomise European public opinion.

Defeatism, however, is utterly foreign to archaic ways of thinking. It will be necessary to restore the archaic frame of mind if we are to survive in the future.

A certain harshness, a resolute frankness, a taste for pride and honour, common sense, pragmatism, a rejection of all non-selective social organisations, an ethic capable of legitimising – if necessary – the use of strength and that will not back down out of dogmatic humanitarianism when faced by the challenges of technological science, an inclusion of warrior virtues and the principles of urgency and inevitable confrontation, a notion of justice whereby it is duties that legitimise rights rather than vice versa, the natural acceptance of an inegalitarian and pluralist organisation of the world (also on an economic level), an aspiration towards collective power, and finally the communitarian ideal: these are some of the virtues of the archaic outlook. They will be essential in the world of tomorrow, which will be marked by bitter confrontations. *A neo-archaic mindset – which is in no way barbaric, as it includes the pre-humanitarian and inegalitarian principle of justice – will be the only one compatible with the character of the approaching century.*

I. Archeofuturism and the question of meaning. What religion? One of the few obvious things about our age, which both traditionalists and modernists agree about, is that Western civilisation has de-spiritualised life, destroying all transcendental values.

The failed attempts at established secular religions, the empty disenchantment created by a civilisation that bases its ultimate legitimacy on the value of exchange and the cult of money, and the self-destruction of

Christianity have engendered a situation that cannot endure. Malraux[55] was right: the Twenty-first century will witness a return to spirituality and religion. Fine, but in what form?

Already, Islam is making inroads through the breach, offering to fill the spiritual void of Europe. Yet this hypothesis – which may well become reality – is dangerous. Because of its extreme dogmatism Islam would risk destroying the creativity and inventiveness of the European soul, its Faustian free spirit. On the other hand, the Machiavellian plans of certain American strategists has led them to encourage the penetration and entrenchment of Islam in Europe in such a way as to induce paralysis. General de Gaulle's words come to mind: 'It is not desirable to see Colombey-les-deux-Églises turn one day into Colombey-les-deux-Mosquées.'[56]

Another answer to de-spiritualisation is gradually emerging. What we have been witnessing for some time now is the rise of 'savage religions', ultimately pagan in nature. This seems to be in accordance with the old European sensitivity: gurus, clairvoyants, astrologers, cults, charismatic groups are multiplying, but a form of Buddhism tinged with Californian colours is also spreading. This solution, however, would lead us to an impasse. In order to gain any credibility and serve a social function, a religion must be organised and structured and possess a central spiritual axis.

As for the secular and political religions modernity has so eagerly embraced – French Republicanism, Soviet Communism, Maoism, Castroism, National Socialism, Fascism, etc. – aside from generally leading to tyranny, they are incapable of 'reconnecting' to religion and mobilising a folk in the long run by providing it with continuing nourishment and an historical reason for survival.

The Archeofuturist answer might be as follows: a neo-medieval, quasi-polytheistic, superstitious and ritualised Christianity for the masses and a pagan agnosticism – a 'religion of philosophers' – for the

---

55 André Malraux (1901-1976) was a famous French author identified with existentialism. He is often attributed with saying, 'The Twenty-first century will be spiritual or it will not be', although this phrase does not appear in any of Malraux's published works. The quote is sometimes given with the word 'mystical', 'religious' or 'ethical' in place of the word 'spiritual'.

56 This quote is reported to have been uttered by de Gaulle in 1959. Colombey-les-deux-Églises, which means 'of the two churches', was de Gaulle's home town. The full quote reads: 'Do you believe that the French body can absorb ten million Muslims, who will perhaps become 20 million tomorrow and 40 million after that? If we allow integration, if all the Arabs and Berbers of Algeria were considered as Frenchmen, what would prevent them from coming and settling on the continent where the standard of living is so much higher? My village would no longer be called Colombey-les-deux-Églises, but Colombey-les-deux-Mosquées!'

A SUBVERSIVE IDEA

elite. The cathedrals are still standing. Should we tolerate that they be turned into museums once and for all? How long will the European episcopate and clergy continue to play a central role in ethnic masochism by encouraging illegal immigration and turning religious rites into para-political litanies?

However things may be, what appears today as an unthinkable fiction *might* become tomorrow's reality. For the catastrophes that await us might lead to a collective mental earthquake.

* * *

It is necessary to reconcile Evola and Marinetti. The new concept of Archeofuturism draws upon the organic, unifying and radical thought of Friedrich Nietzsche and Martin Heidegger: to imagine technological science and the immemorial community of traditional society together – never one without the other; to consider European man, as Raymond Abellio[57] and Jean Parvulesco[58] foresaw, the *deinotatos* ('most daring'), the futurist; and to have a long memory.

Globally, the future calls for a return to ancestral values across the entire Earth.

---

57 Raymond Abellio (1907-1986) was the pen name of Georges Soulès, a French writer on mysticism. He worked for the Vichy government of occupied France and was the secretary general of the Mouvement Social Révolutionnaire, a French fascist party. After the war, he attempted to unite the forces of the far Left and Right in order to create a Eurasian Empire that would stretch from the Atlantic to Japan. For a detailed examination of Abellio, see 'French Visions for a New Europe' by Stephan Chalandon and Philip Coppens, available at *The Occidental Quarterly Online* (www.toqonline.com/2009/12/french-visions-for-a-new-europe/).

58 Jean Parvulesco (1929-) is likewise a French writer on the esoteric who was a friend of Abellio. Frequently referenced by authors on the Right, he further developed Abellio's idea of a Eurasian Empire. For more information, see the essay cited above.

3

Ideologically Dissident Statements

Politically Correct or Politically Chic?

The idea of 'political correctness' is not based on any sincere ethical feelings or even fear of physical repression: it is based on intellectual snobbishness and social cowardice. Actually, it is about what is *politically chic*. The journalists and 'thinkers' of the system are formulating a 'soft' and bourgeois version of the Stalinist mechanism of domination: the risk is no longer ending up in a gulag, but of not being invited to trendy restaurants, of being barred from places that count and from the media, of losing one's appeal in the eyes of beautiful girls, etc. This is the kind of misfortune that befell Jean Baudrillard.[1] Being politically correct is a matter not of ideology but of social acceptance.

The Ruse of Political Correctness

Political correctness operates through 'sham reversal', an extraordinary ruse: one denounces things like 'political correctness' and 'hegemonic thought', while actually being perfectly 'correct' himself; one gives the impression of being politically incorrect – as Jean-François Kahn does – while perfectly adhering to the dominant ideology. In such a way, every actually rebellious thought is neutralised through sham rebellion. The 'politically correct' people hiding behind the mask of political incorrectness must be smoked out – from

---

1 Jean Baudrillard (1929-2007) was a French philosopher and cultural theorist who is regarded as one of the most important postmodernist thinkers. One of his principal ideas is that contemporary reality is made up of concepts and symbols which have no corresponding meaning in the real world, a condition he termed 'hyperreality'.

Benamou[2] to Bourdieu, without forgetting the whole editorial staff of *Charlie Hebdo*.[3]

From Censorship to Distraction

The system only makes use of brutal censorship in very limited areas: it generally resorts to intellectual diversion, i.e., *distraction*, by constantly focusing people's attention on *side issues*. What we are dealing with here is not simply the usual brutalisation of the population via the increasingly sophisticated mass-media apparatus of the society of the spectacle – a veritable 'audiovisual Prozac' – but rather a *concealment of essential political problems* (immigration, pollution, transportation policies, the ageing of the population, the financial crisis of the social budgets expected to occur by 2010, etc.) through a constant discussion of superficial debates on secondary issues: homosexual marriage, PACS ('Civil Pact of Solidarity'),[4] gender equality among political candidates, doping in sports, the legalising of cannabis, etc. The focus on these *insignificant problems* prevents the discussion of urgent and crucial questions. Clearly, this is most convenient for a political class whose members are only interested in furthering their own careers and 'avoid stirring the waters', according to the principle 'after me, the deluge'.[5] Constantinople is under siege and we're debating the gender of angels.

'Consultation' and 'Negotiation': Scourges of Modern Democracy

'Moderate' politicians have come up with a dreadful notion: *consultation*, which is seen as a way of 'modernising democracy', when it is actually a sign that Western liberal democracy is degenerating and committing suicide. Consultation serves as a pretext for inaction: for it blocks all decision-making by reducing it to bastard and minimalist

---

2   Georges-Marc Benamou (1957-) is a Left-wing French journalist and politician. He was one of the founders of SOS Racisme and was a friend of François Mitterrand. He supported Nicolas Sarkozy in the 2007 elections and afterwards served as an advisor to Sarkozy on cultural matters, arousing much controversy in the role.

3   *Charlie Hebdo* is a satirical Left-wing weekly newspaper in France.

4   PACS, or *pacte civil de solidarité*, is a type of civil union in France which is available to same-sex couples as well as traditional couples, although it gives fewer rights than does marriage. Although it was still being debated at the time that Faye was writing, in 1999 it was voted into law.

5   This quote is attributed to King Louis XV of France (1710-1774), who was the last King prior to Louis XVI, who was beheaded during the French Revolution. Louis XV's irresponsible lifestyle and economic policies are widely considered to have contributed to the Revolution, and his remark is regarded as showing that although he was aware that he was causing problems, he wasn't concerned since he knew he wouldn't have to be the one to deal with their consequences.

IDEOLOGICALLY DISSIDENT STATEMENTS

compromises based on a general agreement among various pressure groups and minor trade unions. In a time of emergency such as ours this is a disastrous approach. Through consultation an attempt is made to conceal the fear of action, of risks and responsibilities, and to avoid falling out with the media, shocking minorities active in the field of political correctness, and stirring the anger of unions that cling to their privileges. Most importantly, it represents an attempt to avoid conflicts and problems: the need to face lorry drivers, the 'young', teachers, etc. The catchword here is 'avoid stirring the waters!' To hell with the general interest! Fire fighting is tiring and you could even burn your fingers. Consultation means the shipwreck of the democratic state subject to the rule of law: for those in power forgo the programme ratified by the majority of citizens in order to bargain with non-representative institutions. The true 'consultation' is actually represented by general elections. The primacy of consultation only leads to the maintenance of the status quo, conservatism, laissez faire and political regression.

The soft side of consultation is negotiation. When a legal and legitimate political decision happens to shock or harm a tiny but active minority backed by the media, politicians *give in*, emptying the policy of its content out of fear, laziness, cowardice or discouragement. In such a way, exceptions and privileges replace law, and indecision replaces decision, all because of impotence.

Here are a few examples: illegal immigrants can no longer be expelled; any reform of the diseased national education system is made impossible; every plan to reorganise the social security system fails; a rational transportation policy becomes unfeasible – and so on...

The parliamentary Right is a champion in this field: it has never managed to accept the fact that politics is a battle where it is both essential and inevitable to displease part of the electorate, face interest groups, and suffer the moral rebukes of the Left. Right-wing governments have always been soft. They fear confrontation and do not dare to implement the ideas and programmes by which they came to power, for they feel they have no legitimacy to do so. A Right-wing government would rather avoid displeasing those who voted against it rather than please its own electorate. Winning the favour of the Left is a delight for the Right: like those RPR[6]

---

6   The Rassemblement pour la République, or Rally for the Republic, was a Right-wing political party founded by Jacques Chirac in 1976 which claimed to represent the legacy of Charles de Gaulle. Chirac was President of France at the time Faye was writing. In 2002 the RPR was replaced by the UMP, or Union for a Popular Movement.

MPs who were beside themselves when the Left applauded them and praised their 'modern' spirit and Republican ethics after they had announced– against their own party's will – that they were going to vote in favour of PACS!

Under moral and democratic pretexts, consultation and negotiation give a concrete expression to the disgraceful sagging of democracy and the state based on the rule of law. By rejecting the principles of authority and legitimate decision-making, Western political systems are heading towards failure and self-destruction. Might it be that they are paving the way for the return of autocrats?

Establishing 'Ideologically Liberated Territories' and Creating Meaning

In order to break free from the *ideological cage* the system has shut us up in, it is important to establish ideologically liberated territories. The dominant system is too sure of itself and shows its inefficiency and ineptness when it seeks to exercise censorship. This represents a chance which radical currents of thought should jump at, particularly by addressing the young.

The great weakness of the system is that it believes people are stupid and it seeks to narcotise them or get round them by clumsy means – something which ends up tiring people and proving ineffective. The strategy chosen to contain 'dangerous ideas' has been to *defuse all ideas*, whatever they may be, and – most importantly – sterilising thought and reflection. In the media or in social relations, all that is habitual, banal, predictable, analgesic, futile and effortless, or 'moral', 'positive' and 'nice', is also politically correct. The extraordinary mediatisation of sport is part of this device. Yet this vast ideological emptiness and lack of any values (except for the well-worn ones of hypocritical humanitarianism), the complete lack of seriousness in media talk, the superficiality of 'computer game culture', and the sickening repetition of things devoid of any content, perspectives or depth ultimately engender a form of deficiency.

The future and power belong to those who have things to say and real problems to pose. Simply because these people are more interesting, like novelists telling real stories instead of boring fairytales; because they bring up sore points and address the 'real problems of real people', to quote Margaret Thatcher. Any radical project must make its way past the breach created by this age of absolute conservatism. The young are *waiting for some meaning to bite into*.

Society of the Spectacle and Society of Game

The *society of the spectacle* that Guy Débord denounced in 1967 as a society of alienation – one based not merely on economic exploitation, but also on the continuous use of images and objects, and on the multiplying of simulated experiences through the entertainment industry – has grown far more sophisticated. Not merely because of the boom in audiovisual technology and the Internet, but because in order to better capture people's spirits it has focused on the spectacle of *Game*. From time immemorial, games – mock wars – have been forms of behaviour providing strong psychological release that have fascinated man and allowed the 'masters of the game' to control both actors and spectators. The games in the Roman circus were a political means of loosening tensions. Today we are witnessing the growing influence of games: sport shows followed by billions of people on TV, a boom in video games, TV games and soon virtual games as well (the pinnacle of simulacrums), the multiplying of products offered by 'Française des Jeux'[7] and 'funfairs'... But the game *par excellence* is the sphere of *emptiness*. In games there is nothing at stake *a fortiori*[8] for those who are mere spectators trapped in the *pseudo-mobilisation of their own hubris*. This is a real piece of cake for the system: 'Pay and play, pay and watch others play.' It is hardly surprising that Western states are fostering this society of the game like ancient Rome did during its decline, only in a far more influential way thanks to audiovisual and computer technology.

The CD-Rom games so widespread among the young distract them from dangerous activities like reading and thinking: games do away with those intolerable viruses called ideas.

This strategy adopted by the system, however, seems destined to fail soon. It is the same as the one adopted by Orwell's Big Brother in *Nineteen Eighty-Four* or in the film *Fahrenheit 451*,[9] only in a softer version. A society cannot last long without any sort of positive

---

7   Française des Jeux runs the French national lottery, and also owns betting parlors and on-line games.
8   Latin: 'with stronger reason'.
9   *Fahrenheit 451*, which is based on Ray Bradbury's novel of the same name about a future society in which all books have been banned and firemen burn any books which are found, was adapted into a film by the famous French film director François Truffaut in 1966.

legitimisation. Turning people's attention away from the failures of society by treating them as children – 'Go play and leave Dad alone!' – is a poor and demoralising strategy that won't solve the increasingly serious problems of society. Without mobilising objectives, the ruling ideology will not be able to overcome the distance it has created by relying on emptiness and negativity, and on a culture based on insignificance, the entertainment industry, amusement, and permanent distraction.

The Distortion of Sports

The 'sport gods' of pre-War mythology are dead and gone. On a global scale, sport has not only become an industry (the turnover of FIFA[10] is greater than the state budget of France) – a cause of widespread corruption, doping, and astronomical earnings – but it is also an essential part of showbiz. For this reason, as a new opium for the masses in a West lacking any religion, it fully contributes to the overall lobotomisation of society.

The spectacle of sport infantilises consciences, conceals social problems and the failings of politics. France's success in the last Football World Cup is a sensational example of this. It has been presented as the 'victory of multiracialism and successful integration' and the 'symbol of a France that is finally winning', but this is only mockery, falsehood, and dissimulation.

Here are a few facts: bringing together eleven athletes of different ethnic origins who are paid millions of Francs is a 'borderline case' that is not indicative of any real 'integration' in the general population – integration in a football team does not prove the level of ethnic integration reached by 'pluralist France'; on the contrary, through a sham example it helps conceal the utter failure of the Republican melting pot. While it credited North Africans and Blacks with the victory, it forbade their co-religionists from entering stadiums for 'security reasons'! 'Coloured' supporters, particularly girls, filmed by the cameras with faces painted red, white and blue were seen by the intelligentsia as proof of the fact that 'multiracial France works': what nonsense! As in Brazil, whose multiracial society is actually a multiracist society, the presence of 'coloured' football champions helps conceal reality. As soon as the lights of the sports victory went out, revolts broke out again

10   The Fédération Internationale de Football Association, or International Federation of Association Football, is the international governing body for football, based in Switzerland.

IDEOLOGICALLY DISSIDENT STATEMENTS

in the cities, as did bloody brawls on the streets and in schools. In homage to the naturalised Kabyle[11] football player Zinedine Zidane[12] we have seen rows of Algerian flags waving in the Champs Élysées. After two victories of the French national team, ethnic gangs have repeatedly clashed with the police and British supporters in Paris and Marseilles: what an achievement of 'integration'! A pinnacle of idiocy (and racism) was reached when the newspaper *Libération*, the official organ of conformist anti-racism, criticised the German team because it only marshalled 'blonde players' and no Turkish immigrants or immigrants of other ethnic background on account of the law of blood,[13] thus suggesting that the defeat of Germany was due to its shocking ethnic 'purity'.

To sum up, the victory of a multiracial football team has served to conceal the concrete failure of integration. Far from favouring multiracialism, it has increased multiracism, as shown by the aforementioned incidents that occurred.

Has the victory of the French team contributed to mend the 'social fracture' and fight 'marginalisation'? Has it served to create new jobs or prevent the flight of French brains to California? Has it strengthened the diplomatic and cultural standing of France in the world (with McDonald's as a sponsor of the Championship...)? Has multiethnic society shown itself to be superior to the monoethnic? The answer is no. Sport has simply been prostituted to lend credit to political lies.

The religion of football, the collective hysteria it engenders and psychological disturbances it causes (with supporters going bankrupt to buy tickets from scalpers and touts that cost the equivalent of three months of their salaries) illustrate the deviated function sport has now taken on: to create a lucrative economic sector and mass spectacle resulting in the *manipulation of people's political conscience*. The system focuses the spirit of the masses on ludicrous events; to be more exact, *through sport it turns a neutral spectacle into an event highly charged with meaning.*

---

11 The Kabyle people are ethnic Berbers from the region of Kabylie in northeastern Algeria.
12 Zinedine Zidane (1972-) was the captain of the French national football team that won the 1998 World Cup and the Euro 2000. He retired in 2006.
13 The policy of *jus sanguinis*, or the law of blood, holds that citizenship in a nation is a matter of ancestry rather than the location of one's birthplace. In the German case, this means that a person of German ancestry who is a citizen and resident of another country can become a German citizen, while the children of immigrants who are born in Germany are not eligible to become citizens. Most European states other than France uphold some version of this law.

In such a way, modern sport resumes the role it had in decadent Rome – *'panem et circenses,'*[14] 'RMI[15] and football': it tells lies and makes people forget. Modern sport is run according to the same logic – albeit in a softer version, as we are afraid of blood and what is real – that inspired the producers of the gladiatorial games, where adulated and highly paid slaves would fight against one another.

Sport as Circus

A justification given for the sports spectacle is that it serves to prevent wars by staging symbolic and pacific conflicts, thus neutralising nationalistic drives. Yet the history of football shows that just the opposite is the case, with clashes between supporters and hooligans fuelling nationalistic urges. In Europe the nationalism and chauvinism that would seem destined to disappear are instead nourished by the support of national football teams...

It is easy to note the mental dulling and infantile regression caused by this anger in sports. It is disheartening to see the male population – and now the female too – passionately discuss the performance of a team of players that has no impact on their lives or that of their country. Problems of no substance or importance are thus capturing the attention of the general public.

Sport also nourishes a destructive fascination with brute physical force, which is the opposite of physical courage (that of the soldier) or even 'physical shape' – for the body of great athletes is often damaged by over-training and doping. Society makes up for its lack of physical courage by fawning upon *quantitative physical performances* of no interest. This cult of quantified performance, a by-product of unrestrained materialism – an obsession with who is the fastest, tallest, most muscular and enduring, etc. – finds an expression in the undisputed field of *records*. Athletes who have broken physical records are led to the triumph: a veritable *animalisation* of man – the negation of his intellectual dimension. For ultimately, any hare, greyhound, horse or ostrich will always beat Ben Johnson at sprint racing; any chimpanzee will beat up Tyson, the world heavyweight boxing champion; and

---

14 Latin: 'bread and circuses'. This refers to any policy that relies on keeping citizens happy by distracting them from the realities of social problems.

15 Revenu minimum d'insertion, or Insertion of Minimum Revenue, is a type of French social welfare first introduced by the Socialist Party in 1988, which provides money for those who are not working but are not receiving unemployment payments.

IDEOLOGICALLY DISSIDENT STATEMENTS

as for high-jump records, who can beat the peregrine falcon, with his 5,550 metres?

One may retort that there are sports which require intelligence, skill and courage, like tennis, skiing or sailing. No doubt: but do two guys throwing a ball at each other over a net really deserve all this media attention? Are the performances of trapeze artists or lion-tamers at the circus not equally admirable? And as for extreme sports – transatlantic regattas, the crossing of the Antarctic on foot (when will it be done on one's hands?), or of the Pacific in rowboats – there is an air of pointlessness, boredom and *emptiness* to them. As we no longer know what to do, let us invent something, let us run some (calculated) risks so that sponsors and the media will take notice of us. Once there was a point to the regatta of the four masts on the rum course:[16] to transport this product in the shortest time in such a way as to be the first on the market. Today regattas of this kind are *pointless* performances, meaning they serve no purpose: they are empty tasks –well-paid shows and nothing more; basically, they are global circus events with no laughing clowns.

Curiously enough, the only interesting sports that remain are ethnic ones, which are not mediatised on a global level, like the Basque *pelota*. [17]

Should we then condemn sport? Not if it is understood as *physical exercise* for amateurs, if it serves to improve bodily performances in an intelligent way or to train for combat. In these cases, sport is *targeted*: it has a purpose. The Olympic games of ancient Greece, which today have utterly lost their original meaning, were in no way 'sporting events': they were a form of *military training*. There were no professionals at the Olympics, only amateurs.

The globalised sport spectacle of today has two functions: it stirs false and infantilising enthusiasm for non-events which neutralise people's ideological and political conscience; and they feed a new sector of the entertainment industry which creates very few jobs but is often infiltrated by mafias, while mobilising huge financial resources from which many profit.

And what place does bullfighting have in all this? Well, it is not a sport. It is *bullfighting*.

---

16 A type of boat race.
17 The Basque *pelota* is a game similar to tennis, native to the people of the Basque Country, currently in Spain and France.

The Return to Celebrations

Always according to the circus-games logic, besides sports the system has also encouraged the staging of celebrations: Gay Pride, Technopride, World Music Day, etc. There is nothing spontaneous in these celebration, which do not stem from folk traditions or civil society, as do the various holidays, carnivals, solstices, processions or dances – such as Siena's Palio[18] or Munich's Bierfest[19] – dotted across Europe. These celebrations are consciously and artificially organised and funded by the state, as unstructured outbursts of hubris that serve as collective drugs. They have no *meaning* and in no way embody expressions of popular joy. Besides, these mock celebrations are systematically controlled by the police and end up in riots.

Religious Anathemas and Inquisitorial Thought

In an article published in August 1998 in *Marianne* magazine,[20] Pierre-André Taguieff,[21] an out-and-out yet still ambiguous theorist of 'anti-racism', engaged in an exercise that perfectly illustrates the clumsiness of his current of thought, which dominates the media. With the excuse of exposing the 'dangers' posed by the Front National, he violently attacks the theses of a demographer and economist apparently close to that party, who argues: first, that recent immigrants cost France over 200 billion Francs per year; and second, that the influx of illegal immigrants each year is very substantial. Taguieff presents these claims as fanciful. Yet, nowhere in his article does he base his argument on any scientific facts like figures and statistics; nowhere does he *concretely* refute his opponent. This is rather amazing coming from a thinker who claims to be rational and scientific. Instead of quoting figures and facts – which he does not have, of course – he resorts to *moral accusations of a quasi-religious nature*: he argues that to denounce excessive and costly immigration is to pave the way for 'ethnic cleansing' – in other words, to

---

18 A *palio* is a traditional type of athletic contest between neighborhoods in Italian towns, famously involving horse races. There are many although the most famous one takes place in Siena in Tuscany.

19 Faye undoubtedly means Oktoberfest.

20 *Marianne* is a weekly French news magazine which began publishing in 1997.

21 Taguieff (1946-) is a French sociologist whose work has focused particularly on the issue of racism. In his writings, he has frequently accused the Nouvelle Droite of being racist due to its rejection of the idea of cultural assimilation. In 1994 he published a book on the subject entitled *Sur la Nouvelle Droite*. It is untranslated. Some of his writings on the New Right have also appeared in the American journal *Telos*.

prove guilty of the mortal sin of 'racism', punished by the secular Republican religion.

As the Inquisitors once did with Galileo, *facts* are here answered with anathemas and appeals to a dubious transcendental ethic. What an extraordinary historical reversal: the heirs of Enlightenment rationalism are resorting to irrational and magical or quasi-religious arguments; the heirs of the theories of liberty of expression and emancipation are asking for the banning and criminalisation of the theses (and observations) that upset them; the heir of egalitarian democracy, in the name of 'ethical' and quasi-metaphysical reason, are denying people the right to have their say on the issue of immigration – as well as many others!

Short of arguments, the 'enlightened' elites are using the very weapon they accuse their opponents of resorting to: the obscurantism of tyranny.

On Cinema and American Cultural Hegemony

Like many others, in his last book Godard[22] lamented the domination of American cinema. I have worked for the American cinema industry (in the production of 'French versions' of their films) and have seen what this world is like from the inside. Here are a few concrete facts:

1. American cinema dominates the world market because it sees itself as an industry and not merely a form of 'creativity'. A Hollywood film is not simply a 'work': it is also an advertisement for a whole series of products (consider for instance *Star Wars* or *Jurassic Park 1* and *2*...). The industrial nature of a work does not necessarily deprive it of its artistic value, as people in France believe.

2. The success of Hollywood blockbusters is due to their imaginative and epic character, their dramatic quality, and the ultra-professionalism of the production and its distribution, as well as the perfect technique behind them... This more than makes up for the frequently poor scripts of these movies, with their often childlike and syrupy clichés. Hollywood is like the Jules Verne of film-making, and its scripts are actually often written by Europeans fed up with the lack of dynamism in European productions.

---

22 Jean-Luc Godard (1930-) is an experimental filmmaker best known for his association with the French New Wave in cinema during the 1950s and '60s. He has always been deeply distrustful of Hollywood.

The French and the Europeans (with the exception of Luc Besson)[23] *have lost their taste for epic and fantasy*. What prevents us from regaining it? What forbids it? Why has no European thought of dealing (in our own way, which would no doubt be more intelligent and equally dramatic) with the themes found in *E.T.*, *Jurassic Park*, *Armageddon* and *Deep Impact* (collision with an asteroid), *Twister* (tornados) or *Titanic*? Financial excuses, as we shall see, do not hold water. The same goes for novels, where translations of American thrillers are flooding the market. What prevents us from taking up the tradition of Jules Verne, Paul d'Ivoi[24] and Barjavel?[25] Where are our Philip K. Dicks,[26] Stephen Kings, Robert Ludlums and Michael Crichtons? What we have instead – as happens with cinema – is a literature that ignores and scorns popular genres and produces snobby, boring works focusing on very limited issues, which do not sell well. To implicitly believe that a popular work must be of inferior quality is to betray Molière.[27] In short, American cultural hegemony in regard to films and novels (hence all popular audiovisual entertainment industries) can be explained, despite the often mediocre quality of these products, on the basis of the epic and imaginative character of their themes. The public prefers a highly dramatic work lacking grand ideas and aesthetic perfection to boring but aesthetically and intellectually charged works. The solution for European producers seeking to stand up to Americans is to create works with a highly dramatic, popular character and with scripts of a high cultural level. Our novelists knew how to achieve this in the Nineteenth century.

3. To explain American hegemony in these fields financial reasons are invoked, as well as the presence of the 'vast monolingual American market', which by itself suffices to make productions later exported profitable. But this is sheer sophistry. A blockbuster, promotion

---

23 Luc Besson (1959-) is a French filmmaker who has produced films both in France and the U.S. Among his works is the 1997 science fiction film *The Fifth Element*.

24 Paul d'Ivoi (1856-1915) was one of the first authors of science fiction in France. His work is untranslated.

25 René Barjavel (1911-1985) was a French writer best known for his science fiction. He is credited with inventing the 'grandfather paradox', which speculates about the consequences of a time traveller going back in time and killing his own grandfather before he himself is born, in *Le Voyageur imprudent* (translated as *Future Times Three*).

26 Philip K. Dick (1928-1982) was an American writer best known for his science fiction, which often explored the philosophical problems of reality vs. Illusion. He is widely regarded as having been one of the great American authors of the Twentieth century.

27 Molière (1622-1673) was the stage name of Jean-Baptiste Poquelin, a French playwright who is regarded as one of the masters of comedy.

IDEOLOGICALLY DISSIDENT STATEMENTS

included, will cost 100 million dollars at the most. This is a small business investment that Europeans would be perfectly capable of making. It would be less costly than the *'hôtels de région'*[28] generously funded by our taxes, or the extension of a subway line. Suffice to consider that *Les Amants du Pont-Neuf*[29] – an intellectual and soporific trash-flick that was financed by taxpayers thanks to Jack Lang's[30] lobbying and which was a complete commercial fiasco – cost as much as a Hollywood blockbuster (the neighbourhood of Pont-Neuf near Montpellier was rebuilt in life size)! We might think we were dreaming, but we're not: this is all real. We cannot accuse Americans (as Belmondo[31] does) of 'crushing our cinema'. As for the monolingual American market, it is an argument that does not stand. New technologies have substantially cut down dubbing costs. Films can be shot in any language, knowing that in Europe versions with subtitles will be accepted by the public – something which is not the case in the United States. A French film could easily cover its production costs by being distributed in the non-francophone European market. Provided, of course, it is a popular movie... But few people like the word 'popular': it sounds bad and for critics and decision-makers (usually of the Left) it does not suggest quality.

4. Americans often say that 'the French have amazing talent, but they do not know how to develop it, they are unprofessional' (for 'they practice professional amateurism'). It is true that filmmaking in France lacks rigour; cronyism and nepotism are widespread (the offspring of institutional stars, who are usually not very gifted, tend to usurp the place of young talents); financial structures are loose and unclear; the promotion of movies is badly organised, etc. The same problems can be found in the case of novels. The result is that talent, when it is found,

---

28 These are the buildings for the regional administrations, which have been criticized for being too luxurious.

29 *The Lovers on the Bridge* was a 1991 film directed by Leos Carax. Set around a public bridge in the centre of Paris, the production was forced to build a replica of the bridge in another location when one of the leading actors sustained an injury and filming at the actual bridge went beyond the time allotted. The additional funding provided to allow filming to be completed ended up being insufficient, and production was again shut down until new financiers could be found. After receiving 30 million francs, the film again ran over its budget and was again shut down. It ended up costing another 70 million to finish, with a total cost of well over 100 million francs in total for the entire production.

30 Jack Lang (1939-) is a French politician of the Socialist Party. Between 1988 and 1992 he was Minister of Culture. He currently serves in the National Assembly.

31 Jean-Paul Belmondo (1933-) is a French actor who appeared in many of the most notable films of the French New Wave, particularly Godard's *Breathless*.

is wasted and gifted people often have a harder time finding work than mediocre people with friends in high places or who are part of a clique. This is a French malady that was already denounced by La Fontaine[32] (the courtesan syndrome) and Balzac[33] (the need for reference letters).

Here's an anecdote: in 1995 I met a young French artist who was extremely gifted but could not find a job. He was on the dole and was struggling to get food on his plate. He wasn't part of any clique or mafia; he was Breton,[34] heterosexual, married, and the father of four children. To put it bluntly, he was a rare fellow indeed in Parisianised France. When he offered his services or asked for an appointment over the phone, he never made it past the switchboard operator. So he changed strategy and stopped contacting French companies... Today, he works as an art director in Steven Spielberg's studios in Silicon Valley in San Francisco. This small, gifted Breton, rejected by France, has become a key player in the system of American cultural production, to which he adds his French touch. He is now about to become an American citizen.

Culturally, as well as politically and geopolitically, Americans are strong because we are weak, absent and stiff, and we lack dynamism and will. Let us stop moaning: America is only quite naturally occupying the space we have abandoned.

Social Order and the Pleasure Principle

In societies with well-established values, the 'family' and reproduction of the species, just like the transmission of essential values, are threatened by the emergence of the 'pleasure principle'.

A society founded on order is perfectly capable of integrating parallel practices which only concern a minority of people. This is not a matter of being tolerant or lax, but of adopting an organic approach. On this point both the Right and the Left have been terribly mistaken, for both have adopted a monist logic of exclusion – that of 'either... or' – rather than pluralist values of inclusion – the logic of 'and'. In an organic perspective two opposite principles can co-exist: the fertile

---

32 Jean de La Fontaine (1621-1695) was a Seventeenth century French poet.
33 Honoré de Balzac (1799-1850) was a prolific French novelist, regarded as one of the founders of Realism.
34 The Bretons are an ethnic group native to the Brittany region of France who originally came from Great Britain between the Fourth and Sixth centuries.

and traditional family and deviances, the mother and the prostitute, the serene hearth and the debauchery of the brothel – all within a hierarchical order.

The homosexual lobby and intellectual Left are implicitly attacking the family model and the female role of the housewife, often giving proof of incredible hatred and intolerance. Conservatives, on the other hand, who have a mistaken and fossilised view of 'tradition', always take a puritanical stance.

We should instead revert to an archaic view of things by integrating debauchery and 'orgiasm' – which Michel Maffesoli discussed in *L'ombre de Dionysos*[35] – within the social order. The stronger the latter, the more easily can orgiasm unfold in its shadow, in secret, as was the case in ancient societies. This is simply a wise thing. The 'order principle' is in line with millions of years of laws concerning the reproduction of the species and the transmission of culture and values to one's offspring. The 'pleasure principle' must be tolerated and *hypocritically* managed – for it is human and inextinguishable – yet without allowing it to become the dominating norm and become an order in itself. It should exist subordinately, surrounded by 'social silence'. Does this constitute an apology of lies and hypocrisy? Well, yes. Have you ever seen human societies founded on transparency? Generally, they lead to totalitarianism. Brothels should be reopened.

The less orgiasm is displayed – the more it is virtually simulated through pornography – the stronger it is. The boom in the sex industry is merely a reflection of the sexual poverty of our age. As for adult movies, I have been 'on the other end of the camera' as an actor (why not?). I had a lot of fun, but felt sorry for the poor, frustrated viewers.

I defend orgies, parties and Dionysian pleasures, but only when these are subordinated to the *ordo societatis* (social order) on which they are based. The Bacchanals and Saturnalia of the ancient world... The stronger the social order, the more easily can the pleasure principle and orgiasm unfold in its shadow without being detrimental to social cohesion. Again: the more orgiasm is trivialised, mediatised and displayed, the more intense it becomes. Eros and Dionysus wither

---

35 Michel Maffesoli (1944-) is a French sociologist. The book has been translated as *The Shadow of Dionysus* (Albany: State University of New York Press, 1993). Maffesoli views orgiasm as a healthy thing for modern society, asserting that a 'city, a people, or a more or less limited group of individuals who cannot succeed in expressing collectively their wildness, their madness, and their imaginary, rapidly destructure themselves and, as Spinoza noted, these people merit more than any "the name of solitude"' (p. 8).

when they are shown each night on television. Quality debauchery needs silence and secrecy, i.e., *modesty*, which is the real motor of eroticism and sexual release. But the society of the spectacle and modernity, which invoke emancipation and liberation, are ultimately hostile to libertinism and sensuality, and to all forms of sexual refinement.

Here, as in all other spheres, a return to *sexual joy* and authentic sensuality will only be possible by reinstating archaic principles of order in the context of rigorously ritualised future societies. Archeofuturism...

Homosexuality, the Demographic Crisis and Ethno-masochism

The problem today is that homosexuality tends to surreptitiously impose itself as a *superior model*: a more evolved and suitable alternative to heterosexuality, which is implicitly considered 'outdated'. With the typical intolerance of his pseudo-libertarian current of thought, in a recent column in *Journal du Dimanche* [36] a talented intellectual and writer of the homosexual Left defended the idea of 'civil partnerships' (PACS), claiming he was offended by the fact that the Right denounced these as forms of 'homosexual marriage'. In his hateful and bitter tirade against heterosexual couples, he described families as 'small egoistical cells' ('Le chagrin et la honte',[37] 11 November 1998).

What we are witnessing, then, is a reversal of the previous situation, where homosexuality was abusively repressed. Homosexuality, which ought to have remained within the private sphere, is now imposing itself as a *value* in the public sphere.

There seems to be a disturbing connection between the current demographic crisis, the emasculation of Western societies, and defeatism in the face of immigration and the macho values of Islam on the one hand, and the latent apology of male – and now even female – homosexuality on the other. It is as if, surreptitiously, because of *ethno-masochism*, everything European is being perceived as guilty of reproducing an age-old family, sexual and genetic model.

A few years ago, do-gooders attacked a natalist advertising campaign because it showed blond babies. In other words, European

---

36 *Sunday's Newspaper*, a weekly news magazine.
37 'Grief and Shame'.

IDEOLOGICALLY DISSIDENT STATEMENTS 107

natalism is considered a form of racism – being oneself is an attack on others. Fertile European families are seen as guilty of *biological imperialism*. This is an incredible semantic reversal, typical of tyrannical and totalitarian ways of thinking.

It is not a matter of advocating any *repression* of homosexuality, of banning homosexual couples or socially penalising gay people; simply, the prospect of legalising of a form of 'marriage' for homosexuals would have a highly destructive symbolic value.

Why? Whether gay unions go 'against nature' or not is beside the point. Nobody cares about this – it is an endless, pseudo-scientific debate. The fact remains that marriage and legal heterosexual unions enjoy forms of protection and public benefits that are accorded to couples capable of having children, and hence of renewing the generations and thus being of objective 'service' to society. Legalising homosexual unions and awarding them financial privileges means protecting sterile unions. To put heterosexual couples, which perpetuate the population, and homosexual couples (whether male or female) on the same level is a sign of the pathological exasperation of individualism. *It means mistaking desires for rights.* It means scorning the collective interest and riding roughshod over common sense, a notion with which the French Left – the most stupid Left in the world – has been in conflict ever since 1789 thanks to its ideological hallucinations.

To legalise homosexual unions is to plunge into the general confusion denounced by Alain de Benoist, whereby 'everything is equal to everything else'. Why not legalise marriages between human beings and chimpanzees, then? After all, if what matters are individual rights and desires, which is to say personal fancies with no regard for age-old bio-social realities... Progressivism is a form of infantilism.

Besides, gay couples usually don't last long and don't work very well. This is nothing strange: you cannot get away with defying the laws of nature – there's a price to pay for all biological and ethnological anomalies. Let gays live their lives, be tolerated and respected; but let them not impose their norms like a tyrannical minority and claim privileges. As many psychoanalysts have observed – most notably Tony Anatrella,[38] who has reformulated Freud's theses on the matter – homosexuality is a *neurosis caused by immaturity*. Increasing numbers of biologists believe it is simply a hereditary mental disease. Basically,

---

38 Tony Antrella (1941-) has never been translated into English. He continues to advance the thesis that homosexuality is a psychological aberration that requires treatment.

homosexuals, whether male or female, are not emotionally happy. They suffer from their psychosexual illness and feel frustrated because they are incapable of conforming to socio-biological normality and balance.

Homosexuality today is a psychoanalytical problem. Like all minorities that have received some satisfaction and form of acknowledgment, homosexuals are furious that they are no longer the victims: *they feel frustrated because they are no longer persecuted*. They know there is much talk made about them but want more and more. They wish to make up for the disfavours made to them in the past by claiming infantile privileges – hence their aggressiveness, as a counterpart to their inner discomfort.

Having said this, let homosexual unions be legalised, with all the fiscal and matrimonial advantages this entails. As always, it will be the power of reality that will demolish this utopia. *Sic transit gloria imbecillorum*.[39]

The Primacy of Desire over Law

'*Sans-papiers*', illegal immigrants who infringe democratic laws, are allowed to remain in France thanks to the power of the media and minority pressure groups. Their desire thus prevails over the laws voted for by the French people.

Herein lies one of the paradoxes of the ideology of human rights: well-defended private interest overcomes the will of the majority. This opens the door to all mafias...

Lorry drivers, fishermen, pilots, the trade unions of teachers and students (an active minority), subsidised farmers and railwaymen all challenge the law with impunity and force the government to back down in order to defend their egoistical group interests. The media join in, and out of cowardice or careerism politicians give in and don't move a finger.

Everywhere minority interests prevail over the law. What a paradox: the champions of the 'Republic' are signing off on the overthrow of the state subject to the rule of law. They do not realise that an end will be put to these disorders by adopting an archaic but very effective solution: *tyranny*, where the will of a tyrant takes the place of the failing legal system and the will of the majority, yet without yielding to private desires.

---

39 Latin: 'Thus passes the glory of imbeciles.'

IDEOLOGICALLY DISSIDENT STATEMENTS

The above idea is probably shared by Jean-Pierre Chevènement[40] – but most likely nobody else.

The 'Biolithic Revolution' and the Great Ethical Crisis of the Twenty-first century

A conflict will inevitably break out in the Twenty-first century between the great monotheistic religions (Islam, Christianity, Judaism and the secular religion of human rights) and the progress of technological science in the fields of computers and biology. In his book *La revolution biolitique* (Albin Michel, 1998), Hervé Kempf[41] argues that science is about to undergo a 'transition' comparable to the Neolithic Revolution, when *homo sapiens* passed from the hunter-gatherer stage to farming and agriculture, thus shaping his environment. We are now experiencing a second great change, in both biology and informatics. This revolution consists in the artificial transformation of living creatures, in the humanisation of machines (the quantic and especially biotronic computers of the future), and consequently in the interactions between men and robots.

Anthropocentrism and the unifying notion of 'human life' as a value in itself, which constitute the central dogmas of both monotheistic religions and the egalitarian ideologies of modernity, are entering into a sharp contrast with the possibilities offered by technological science, and particularly the 'infernal' alliance between informatics and biology. There will be a major conflict between researchers on the one hand and political and religious leaders on the other, who seek to censor and limit the use of new scientific discoveries – although they may not succeed...

Things such as artificial births in incubators; intelligent, 'quasi-sensitive' and quasi-human biotronic robots; chimeras (crossbreeds between humans and animals, a patent for which has already been filed in the United States); genetic manipulations or 'transgenic humans'; new artificial organs that increase the faculties tenfold; the creation of hyper-endowed and ultra-resistant individuals through

---

40 Chevènement (see also chapter 1, note 15) was initially a member of the Socialist Party who left it in 1993 due to his opposition to the Persian Gulf War and the Maastricht Treaty which created the Euro. He then founded the Citizens' Movement. He was Minister of the Interior in 1998. He is known for frequently resigning for ideological reasons, a rare phenomenon in French politics, and for his strong opposition to the European Union. He is currently a Senator.

41 Kempf (1957-) is a writer on science who has been the Environmental Editor for *Le Monde* since 1998.

positive eugenics; and cloning – all risk shattering the old egalitarian and religious idea of man even more than Darwin and evolutionary theories have done. 'Human factories' are already being developed through the production of artificial organs, assisted procreation, function stimulation, etc. The creation of machines based on biological processes (e.g., neural computers and DNA microchips) is also not far away. The very definitions of man, living creatures and machines will have to be reformulated. Artificial humans and animal machines...

In the Twenty-first century, man will no longer be what he used to be. This will bring ethical confusion with devastating effects. There is the risk of witnessing a mental shock, a clash with unforeseeable consequences between two worlds: the new biotronic or biolithic view on the one hand, and that of the old ideas promoted by the great world religions and the modern egalitarian philosophy of human rights on the other.

Only a neo-archaic outlook will enable us to cope with this shock, because once – whether among the Incas, Tibetans, Greeks or Egyptians – it wasn't man who was at the centre of the world but *deities*, who could take any living form they wished. The technological science of the future invites us not to dehumanise man, but to stop divinising him. Does this mean the end of *humanism*? It certainly does.

Genetics and Inegalitarianism

One of the central theses behind the idea of 'Archeofuturism' is the following: paradoxically, *Twenty-first century technological science is driving modernity's back against the wall*, for it 'risks' rehabilitating inegalitarian and archaic worldviews. Here is a simple example in the field of genetics: the drawing of a 'map of the human genome', the study of hereditary diseases, the development of genetic therapies, research into brain chemistry, AIDS and viral illnesses, etc., are already starting to concretely reveal the *inequality among humans*. The scientific community is caught in a vice: how to obey the censorship of political correctness, giving in to the intellectual terrorism of egalitarianism while at the same time promoting scientific truths that may be therapeutically useful? A conflict will arise here, and a serious one too. Geneticists, sexologists and virologists are already finding it harder and harder to conceal the fact that one of the canonical myths of the religion of human rights – the principle of equality among population groups and the genetic individualisation of humans – is scientifically untenable.

On the other hand, it is clear that biotechnologies (e.g., assisted procreation, biotronic implants, artificial organs, cloning, genetic therapies, and the manipulation of transmissible genomes – technologies which are genuine forms of eugenics, although few would dare use this word) will neither be available to everyone nor covered by social security; moreover, they will only be applied in the great industrial nations. What is *de facto* a kind of eugenics will be offered to a minority which will witness an increase in its life expectancy: the height of inegalitarianism has crept like a virus into the heart of modern egalitarian civilisation. Another embarrassing problem: how will anthropocentric humanists react when chimeras (man-animal hybrids) are created to be used as organ and blood banks or to engineer better semen or to test drugs? Will they seek to ban them? If so, they will fail. To face the global shock of future genetics we will have to adopt an archaic outlook.

The Notion of 'Love': One of the Pathologies of Civilisation

Western civilisation began to grow considerably weaker from the day it started assigning an absolute value to a pathological feeling: *love*. This pathology has eroded both our demographic resources and our defensive instincts. It is a secularised Christian inheritance. Does this mean that *hate* must be the motor of conquering and creative civilisations? No. It is 'love', whether personal or collective, that represents a pathological and emphatic form of solidarity leading to failure and, paradoxically, hate and massacres. Religious wars and contemporary forms of fanaticism on the part of the monotheistic religions of love and mercy are proof of this. Even totalitarian Communism was founded on the idea of 'love of the people'.

It is necessary to have (temporary) *allies* among nations, not friends. Among individuals, it is better to say 'I am fond of you' than 'I love you', and to engage with others according to the logic of alliance rather than the blind – and shifting – gratuitousness of love.

Love is absolute, hence totalitarian. Human feelings and strategies are changeable. Both in politics and in our personal relations, instead of the verb 'to love' let us use adopt a polytheistic range of verbs: to be fond of, admire, ally oneself with, come to terms with, protect, help, cherish, desire, etc. We should not have children as a gift we wish to bestow on the partner we love, but rather because we feel this person is worthy of breeding and perpetuating our stock. Today half of all marriages fail because they are based on an adolescent and ephemeral feeling that vanishes with the first gust of wind. Lasting marriages are based on plans.

The same is happening with parents' education of their children. This is also failing because it is based on the blissful adulation of one's offspring (these by-products of love), which undermines the legitimacy and authority of parents, perceived as loving sheep. Politicians are similarly doomed to failure because their ideology and actions are marred by residues of love – good feelings, do-goodism, humanitarianism, pity, masochism, and a misdirected and hypocritical altruism – instead of resting on the decision-making will of pursuing one's goal to the very end, whatever the cost.

This civilisation, which has long been implicitly founded on the distorted notion of love, must one day return to the allegory of Don Juan, the symbol of anti-love par excellence. Archeofuturism.

Philosophical Debacle and Frauds

The absence of genuine philosophical values in this *fin de siècle*[42] of ours is illustrated by the popularity media commentators enjoy who promote hollow ideas, state the obvious and embrace hegemonic thought – people like Comte-Sponville, Ferry, Bernard-Henry Lévy, and Serres. Do-goodism with no metaphysics or spirituality, petty materialism, a childish return to the Enlightenment, hypocritical moralising and altruism, ethical virtuousness, ethno-masochism, xenophilia, kindness with ulterior motives, and irresponsible humanitarianism: all these mental attitudes are profoundly unsuited to our age. These weakening, emasculating and morally disarming values are misleading in a world that is growing increasingly harsh and which calls for just the opposite: for combative values. While we need a *new philosophy of action*, in this society of bovines we are fed the stale

---

42 French: 'end of the century'. In addition to its literal meaning, it also has connotations of belonging to an age of degeneracy and opulence that comes about as one age is ending, just prior to the birth of a new age. Much of French literature and art from the Nineteenth century, particularly of the Symbolist movement, is referred to by this name.

remains of the Eighteenth century *philosophy of compassion*, passed off as brilliant novelties and advancements of the spirit.

A neo-dogmatic philosophy, capable only of 'communicating' (propaganda), is masked as one of anti-dogmatism, freedom and emancipation, while it is only the academic harping on about obsolete ideas and a tool of intellectual terrorism.

The philosophy of the hegemonic French intelligentsia of the Twentieth century will be remembered for its *plagiarism* (Sartre, Lévy), *pathological altruism* (Lévinas)[43] and *fraud* (Lacan and the structuralists).[44] It stands out for its use of abstruse language to convey 'non-ideas'. This is why the excellent critical work on French philosophy written by Sokal[45] and Bricmont, *Intellectual Impostures*,[46] has stirred such controversy. Only the truth hurts.

To face the future, what we need is an inegalitarian philosophy of will to power. We must return to Nietzsche again, this visionary against the Enlightenment. The revolution that is approaching calls for a new epistemology capable of doing away with traditional humanism. Through a return to archaic values we must envisage man no longer as a divine *being* removed from animal nature but rather as both an *actor* and *material*, someone who experiments on himself according to a Faustian logic.

The Emasculation Process

Advertising follows rather than starts social trends. Its aim is to sell, not spread new fads or ideas. In this respect, all advertising accurately reflects its age, for it is forced to be business-like, effective and capable

---

43 Emmanuel Lévinas (1906-1995) was a Lithuanian Jewish-born philosopher who had a large impact on the development of postmodern philosophy. Among his ideas, he held that ethical responsibility must come prior to the attempt to understand the truth.

44 Jacques Lacan (1901-1981) was a French psychoanalyst whose work had an immense impact on structuralism, and then postmodern philosophy. He frequently cited scientific and mathematical evidence to support his ideas, although professional scientists and mathematicians have said that his evidence is nonsense.

45 Alan Sokal (1955-) is an American physicist who is infamous for having submitted a paper to the postmodernist journal *Social Text* in which he claimed that recent discoveries in quantum physics proved aspects of postmodern philosophy. The science he used to support this was bogus, but Sokal believed that the editors would print it without verifying it since it flattered their preconceptions. Sure enough, the article was published in 1996, and Sokal exposed the hoax after it appeared. This generated a great deal of debate over the value of postmodernism, its attempted use of science to support its theories, and the state of the humanities in general.

46 Alan Sokal and Jean Bricmont, *Intellectual Impostures: Postmodern Philosophers' Misuse of Science* (London: Profile, 1998) was the first of a series of books that Sokal wrote about the controversy. It was published in the U.S. under the title *Fashionable Nonsense*.

of precisely echoing the spirit of the ruling ideology. In a popular mainstream magazine, an ad for luxury shoes set in the changing room of a gym shows a woman facing two naked Black athletes taking a shower, whom she is 'dominating'. The opposite page has a T-shirt ad. The models, two European males, display an effeminate and markedly gay look. They come across as arty fairies with a languid, tired gaze. Find the mistake.

The Responsibility Principle

This is not a conspiracy, it's something worse. It's a kind of 'logic' – a form of collective resignation. Conspiracy theorists are wrong. A strong folk will not let itself be captured or crushed by the system by which it is ruled. All peoples are responsible for their own destiny. What we get is our own fault, not that of others. We are the actors and guilty of our own defeats. A folk is never the passive victim of its own cultural or ethnic effacement: it is its author and an accomplice to it out of resignation and an unwillingness to defend itself. The cultural hegemony of the United States and the gradual and veiled colonisation of France and Europe by the Third World are not merely the product of manipulation. We have let such things happen to us. Our folk had the means – the democratic means – to defend itself. But it chose not to. A 'secret orchestrator' has little power when faced by a folk determined to resist it with all its might.

Archeofuturist Suggestions on Art

Alain de Benoist's magazine *Krisis* has dared launch a *debate* on whether 'contemporary art' is not in fact a kind of fraud. The media have immediately joined forces to denounce this crime of treason on the part of the 'far Right'. Actually, everybody knows – although no one dares say so explicitly – that for almost fifty years now the 'contemporary art' supported by state subsidies and the media has amounted to nothing but academics (and snobbishness), and that it is now gradually collapsing. What a paradox: contemporary art, which through its creative power and vitality was intended to serve as a war machine against academia, is now drowning in the worst conservatism. It shares the destiny of Communism. It has turned into official art, art zero.

The reasons for this are well-known: fraud and incompetence. In the early 1900s, an aesthetic ideology took hold that immediately gave

IDEOLOGICALLY DISSIDENT STATEMENTS

its fruits: artists' inspiration – their message – came to be regarded as more important than their technique and professional skill; knowledge of artistic rules and canons were seen as a form of 'oppression'. Such was the myth of 'the freedom of the artist'. Later, false inspiration gradually took hold: artists, lacking real inspiration and competence, achieved subsidised media success thanks to their connections – as was the case with Calder, Saint-Phalle and César among many others. Artists even stopped trying to 'shock the bourgeois': they only sought to prove themselves to be progressive and started repeating themselves over and over. By then, they had turned into subsidised finger painters. Recently, pieces of graffiti made by mentally handicapped children have been considered 'masterpieces'. I personally devised the following hoax for the *Echo des Savanes*:[47] I painted some canvases before a clerk of the court consisting of daubs vaguely representing phalluses, one minute for each painting... These were then sold in a prestigious art gallery in Rue de Seine to stars of showbiz who enthusiastically admired them. Hoaxes of this sort had already been performed by negotiating a high price for 'canvases' that had been 'painted' by a donkey with its tail (*Sunset on the Adriatic*)[48] and by a female orangutan.

Contemporary art has done away with the crucial notion of *talent*.

Today, in the public sphere, a repetitive and far from creative sort of contemporary art based on sheer fraud coexists with a museum-centred worship of masterpieces from the past. This is typical of societies caught in an aesthetic deadlock. It is interesting to note that the system reacts to all criticism regarding the authenticity and quality of contemporary art with its anathema: 'So you're a fascist!' This is proof of the fact that the system is perfectly aware of the worthlessness of the 'artistic' production it champions and of the burning failure of its politico-aesthetic model. As soon as this sore point is brought up, the system reacts with insults and threats.

Even today, however, creative artists exist who shun the pretentious vacuity of official art: Vivenza[49] with his *bruit*, for instance; the sculptor Michel de Souzy; or painters like Frédérique Deleuze,[50] Olivier

---

47 *The Echo of the Savannah* is a comics magazine for adults.
48 This hoax was perpetrated by the writer Roland Dorgelès in 1910.
49 Jean-Marc Vivenza (1957-) is a philosopher and musicologist with an esoteric bent, having studied Guénon. His *bruit* is a form of industrial music, called noise music in English.
50 Deleuze maintains a Web site of her work at www.deleuze-peintre.com.

Carré, and Tillenon.⁵¹ There are many such artists, but they are viewed with suspicion and alienated because they renew art through the principles of *European aesthetics*: by reconciling aesthetic canons and creative daring, meaning and beauty, technical work and inspiration.

Official contemporary art (which should not be confused with 'today's artists', who are often very talented but silenced) is closely tied to the system. Its aim is to cut the thread and break the lineage of the ascending trajectory of European art. It is always the same will to cultural iconoclasm in the attempt to strip Europeans of their historical memory and identity.

The tactic adopted is a clever one: on the one hand, insignificant works are promoted in the media, usually the non-works of a nobody (after all, in the confused scenario of egalitarianism, where 'everything is the same as everything else', what is worthless can aspire to become art – the fouler and dirtier it is, the more worthy of admiration); on the other hand, a museum-like admiration of the past is elicited – of a fossilised and neutralised past – as a clever way of promoting sterile traditionalism. What matters is for the masterpieces of the past not to serve as a means for reawakening talent in the present or future. The aim is to destroy European artistic creativity, with its magnificence, aesthetic power and talent; to corrupt peoples' taste by presenting the works of talentless individuals as works of genius; and to do away with all traces of any European aesthetic personality by severing the cultural roots of art. Such has been the often unconscious but always implicit strategy of the 'masters of art' for several decades now. This strategy reflects a form of *envy* (a feeling that, along with desire for vengeance and resentment – as Nietzsche understood well – has always played an important role in politics and history): envy of and resentment against the innate talent of European art.

Part of this enterprise is the ridiculous cult of 'primitive arts', of which a naive man like Chirac has become a promoter. A primitive statue is considered as good as Michelangelo's *Pietà* – isn't that so, Mr. Pécuchet?⁵² Here, too, egalitarianism clashes with common sense and reality, condemning itself.

---

51 Yann-Ber Tillenon was part of GRECE but left at the same time as Faye. He remains active in the Right alongside Faye.

52 Pécuchet is a character in a novel by Gustav Flaubert, published in 1881: *Bouvard et Pécuchet*. The two title characters are office clerks who become friends and, out of their shared enthusiasm for learning, attempt to master all of the various branches of knowledge. All of their efforts are unsuccessful.

IDEOLOGICALLY DISSIDENT STATEMENTS 117

Genuine, unrepressed aesthetic creation has sought refuge in *technique* through an unconscious return to the Greek tradition of aesthetics as *technè*[53] and *chréma* [54](objective usefulness). It is the designers of cars, planes and objects who are producing artworks today. What do we prefer? A crushed Renault by that fraudster César[55] or a Ferrari signed by Pininfarina?[56] It may well be that people will soon grow tired of the false masters of official art – this has already begun to happen with the decline of the FIAC (Foire International d'Art Contemporain).[57]

Bourdieu, or the Impostor

Bourdieu denounces the bombardment of TV[58] but reflects its ideology in his own thought. He is the self-proclaimed *maître-à-penser*[59] of the 'Left of the Left', which is to say of the New Left, without ever proposing any credible solution to the ultra-liberalism he sees as all-pervasive. Still, he doesn't mind having his photograph taken for the media and to appear on that very same television he claims to hate. Bernard-Henry Lévy and Mgr. Gaillot[60] mustn't be all that keen on this media dinosaur. He's a funny character, Bourdieu...

He once flirted with the Nouvelle Droite, in the early '80s, when it was quite fashionable. We would have lunch together at the *Closerie des Lilas* and discuss Nietzsche and the reversal of values. It was the anti-liberalism of the Nouvelle Droite that attracted him. But like all those of his kind – Parisian intellectual bureaucrats – Bourdieu wasn't really interested in ideas. He was more interested in himself. Tragically lacking any theories, the new intellectual guru of a vaguely resuscitated far Left was only capable of countering the 'hegemonic thinking' of ultra-liberalism with another hegemonic idea: an outdated reissue of Marxist conservatism. Like the whole far Left, Bourdieu is incapable of formulating any analysis

---

53 *Technè* is the method involved in creating an object or in accomplishing a goal.
54 Greek: 'thing'.
55 César Manrique (1919-1992) was a Spanish artist. Among his works were collections of compressed car parts, including some from Renaults. He died in a car accident.
56 Andrea Pininfarina (1957-2008) ran an Italian car body firm of the same name. He died in a car accident.
57 International Fair of Contemporary Art. Paris has been hosting this fair every October since 1974.
58 Especially in his book *On Television and Journalism* (London: Pluto Press, 1998).
59 'Master for thinking'. Less literally, it implies any teacher from whom one develops a particular way of thinking.
60 The Most Reverend Dr. Jacques Jean Edmond Georges Monseigneur Gaillot (1935-) is a former Catholic bishop nicknamed 'The Red Cleric' because of his extreme Leftist positions. He was removed from his position by the Vatican in 1995 for publicly opposing several of the Church's precepts.

pertinent to the present social situation. Like many others, he illustrates the shipwreck of Leftist intellectuals. After having fooled themselves *with* their ideas, they are now sinking *without* any ideas.

The Method of Dependence

The tamers of tigers and other wild beasts do not use brutal methods such as beatings, punishment and privations to subdue their animals into submission. It would be too dangerous and complicated. The winning strategy is the carrot, not the stick. The animals become *dependent upon useless but enjoyable rewards*: sweet food or protein, petting, sexual favours, etc., after each act of obedience, so that their ability to rebel against their masters is weakened or annulled.

The ruling system and ideology make use of a refined version of this method. They no longer send dissident citizens to labour camps – this method is outdated. Rebellion is now put to sleep and marginalised, not only by directing people's attention towards irrelevant things (the football World Cup, etc.) through the classic strategy of intellectual stupefaction, but by adopting the method of *dependence*. The system makes civil society dependent by assigning rewards, advantages, privileges and useless gadgets.

Like those given to caged wild animals, these are false advantages. We are led to believe that we are free when we are in fact prisoners, that we can move around faster on the grand tourers that cost us a fortune when we have to spend hours caught in traffic or at work to pay for them. We are dependent on the holidays we have to plan, on our TV fix, and on an 'unrestrained desire for useless objects', as Baudrillard has observed. This is a soft dictatorship, intended to make us forget about unemployment, job uncertainties, adulterated food, environmental degradation, and the gradual extinction of our folk. We are living in cages like animals in the zoo but are *physiologically* happy. We are Nietzsche's 'last men', who gleefully thank their masters.

The Reign of *Arnaque*:
False Transparency and Forgeries

In argot,[61] the word *arnaque* is used to describe a kind of 'soft swindle'. The yellow line of actual swindle is not crossed but only touched. It is

---

61 Argot is any language which is used by a particular subculture, such as criminals, in order to prevent outsiders from understanding what they're saying.

IDEOLOGICALLY DISSIDENT STATEMENTS

like failing to stop not at a red light but at a dark orange one. It is a sign of our times that, once chiefly confined to companies found guilty of 'false advertising', *arnaque* has now become one of the chief motors of advertising and the consumer's drive. Today it is practiced by all businesses and reputable companies, and even by the state. So much for the theory – here are a few examples.

Competing companies will reach a mutual agreement (the method of oligopoly) whereby they will produce short-lasting products that 'must' soon be replaced: car bodies that become rusty in under three years, components of audiovisual devices that break down after 500 hours of use, fridge compressors that give up the ghost after four years, jeans that become torn after twenty washings, etc.

A 'culture of *arnaque*' has taken hold to which the state is largely contributing. A patent illustration of this: while experts had solemnly declared and sought to prove that in 1998 there would be a decrease in direct taxation and compulsory charges, just the opposite has happened: there has been an increase, making the economically disastrous fiscal and nationalisation policies of the state even worse.

The other side of *arnaque* and deception is *false transparency*. People insist that they are being honest and concealing nothing both in politics and in agribusiness. This helps to establish false confidence. A few examples: food producers generally conform to the law that forces them to state in the case of each product whether it contains things such as emulsifiers, flavour enhancers, colours or thickeners. On the other hand, what few people know is that while the law has allowed the use of these additives – because of pressure from the agribusiness lobby – 50% of these substances have been found to be carcinogenic in lab animals; they probably are for humans too, if consumed on a regular basis. Yet, false transparency – the 'There's nothing I'm concealing' approach – engenders suspicion. Only half the truth is spoken. 'Yes, I do put E211 in the tomato sauce you buy', says the producer; and because he admits it, you believe it isn't toxic when it actually is.

The media and television are the realm of deception and special effects: false live broadcasts, organised exchanges of favours, deceptive advertisements, the promotion of friends or people towards whom one is in debt, the rejection of all *critique* (whether cinematographic or literary), etc. Spontaneous 'talk shows' are actually produced like dramas with an official message to convey. The present audiovisual system leaves no room for spontaneity and truth, although it invokes these as its source of legitimisation. It can be stated without any exaggeration that news broadcasts today are far more censored, manipulated and counterfeited – and with far greater skill – than they were at the time of the ORTF[62] under de Gaulle. Patrick Poivre d'Arvor[63] is nothing but a puppet, as are the people of the Canal Plus[64] puppet show that represent him.

Arnaques and deception: these are no longer practiced by small-time fraudsters alone: with amazing cynicism, they are also practiced by mainstream public and private institutions under the redundant banner of *transparency*. As explained by Primogine (the author, with Thom, of catastrophe theory), when a system gets to the point of justifying *a* through *non-a*, it is on the verge of collapse.

The Logic of Hypocrisy: The Dialectic of Spoken and Practiced Morals

Moral discourse has never been as exacting and rigorous as it is now. The system and its media preach against violence, racism, and chauvinism, for the rights of everyone, goodness, kindness, independent justice, universal love, equality, social justice, democracy, and 'civil conscience'. A sermon worthy of a pious old lady.

Reality, however, is radically different: political corruption, the collapse of social rights, the toleration of urban violence as well as that shown by the media, an increase in economic disparities and injustice (Leftist billionaires are the first to discuss social justice), the disappearance of solidarity among close people in the face of individualistic egoism, impunity for groups breaking the law, privileges accorded to professional categories that already enjoy protection, a growth of precarious jobs exploited by the public sector, etc.

Things have always been so. Psychiatrists call it the 'compensation effect': the more a social system is defective, the more its discourse is aimed at praising the qualities it lacks. Immoral people speak in moralising tones. This is not merely a form of exorcism, but an attempt to make people forget: 'They shouldn't realise what is happening.'

---

62 Office de Radiodiffusion Télévision Française, which was the agency which ran public radio and television in France between 1964 and 1974.

63 Poivre was a well-known TV journalist in France for more than 30 years. He was fired in 2008 after defamation charges were filed against him.

64 Canal+ is a French pay television channel.

IDEOLOGICALLY DISSIDENT STATEMENTS

The central weakness of the system – and the ruling ideology – is that it cannot continue to lie for long. As U.S. Senator Gingrich[65] put it, 'You can lie to a woman ten times and once to a nation, but you can't lie ten times to a nation.' In the long run, the absence of concrete results in the project for a global society cannot be concealed by means of empty countermeasures: intellectual stupefaction, the turning away of people's attention, the numbing of minds, and dependence. *Concrete reality* is backfiring. People are asking for results, as despondency has its limits – and these are imposed by tangible facts: the lies regarding the fall in unemployment, economic uncertainty and anxieties, an increase in the poor despite the growth registered, an objective increase in insecurity despite all falsified statistics, immigration making its presence felt more and more, etc. Even the highly effective propaganda on TV, which seeks to give the impression that 'all is going well' and tries to demonise and criminalise those holding opposite opinions, will meet its end sooner or later. When the lion no longer has anything to eat, it eats its tamer. The lion in this case is the people.

Negative Legitimisation: The Tale of the Big Bad Wolf

Western democracies are failing to implement their utopia and so are denouncing an imaginary enemy. Politicians no longer say, 'Vote for us, because we've got the right solutions and we'll improve your living conditions because our solutions are the best.' This is positive legitimisation. Politicians now are instead – implicitly – saying, 'Vote for us, since even though we're a bunch of good-for-nothings, bunglers and bullies, this is nothing serious: at least we can protect you against the return of Fascism. If it weren't for us, you wouldn't even have eyes left to cry....' This is negative legitimisation. The redundant commemorations of Second World War events and the voyeuristic descriptions of 'Nazi crimes' with trials and denunciations which are being incessantly broadcast on the media over fifty years after they took place are all part of this strategy.

This is the big bad wolf tactic: 'Daddy is bad but if you do not obey him, the big bad wolf will come and get you. And that will be even

---

65 Newt Gingrich (1943-), a U.S. Republican Congressman from Georgia who is best-known for his role in leading the so-called 'Republican Revolution' in the 1994 mid-term election, and for his role as Speaker of the House of Representatives from 1995 until 1999. He is still a noted conservative commentator in the U.S.

worse!' The system, which is failing to gain consensus and achieve any results, is inventing virtual enemies which it claims to be protecting the people against: 'The Front National is the NSDAP[66] under a new guise; if we expel too many immigrants, there will be an economic collapse and a dictatorship will be installed.' This old strategy has its limits and these will soon be evident.

The 'Republican Front': The Antechamber to the Single Party

'Front Républicain'[67] vs. 'Front National': this is the current call-and-response in the world of politics. The Republican Front, which fancies itself to be the guardian of pure democracy against the 'fascist threat', is actually the product of a far-Leftist and para-Trotskyist minority whose tradition for the past seventy years has been totalitarianism. The fight against the Front National reveals the unbearable contradictions behind this Republican Front so intent on saving democracy: for it is neither Republican nor democratic. How can this be doubted? When a society excessively appeals to a given political notion (for instance, democracy or citizenship), it means this very notion is in peril. Pseudo-democratic emphasis serves to cover up a regime that is growing less and less democratic. The discourse of the Republican Front takes up the rhetoric – which is actually totalitarian – of the Convention of 1793 – of the fathers of the Reign of Terror.

At a 'spontaneous' demonstration in Lyon against the supposed alliance between Charles Millon[68] and the Front National, Louis Mermaz[69] explained that it was a matter of 'fighting against the unacceptable: the Front National co-administrating a region'. So according to this 'democrat' it's 'unacceptable' for regional councillors that have been democratically elected by the people to fulfil their office. This slip on Mermaz's part means: democracy is not open to all; or rather, it is *unacceptable* that democracy may play by all of its rules; or again: it is

---

66 Nationalsozialistische Deutsche Arbeiterpartei, the National Socialist or Nazi Party.

67 The Republican Front was a coalition of both Left- and Right-wing parties put together with the express purpose of keeping the Front National out of power.

68 Charles Millon (1945-) was a member of the centrist Union for French Democracy (UDF), and served as Minister of Defence from 1995 until 1997. He was also the President of the Rhône-Alpes Regional Council. In 1998, faced with defeat as President in the election, he agreed to accept the votes of the Front National, although this led to him being expelled from the UDF. He currently holds no office.

69 Louis Mermaz (1931-) is a politician of the Socialist Party.

IDEOLOGICALLY DISSIDENT STATEMENTS

unacceptable according to our limited vision of democracy that voters may vote for someone other than us, the Republican Front.

This Republican Front includes: 1. the Communist Party (PC) and the far Left; 2. the Greens and the Socialist Party (PS); 3. a 'Republican Right' that is emasculated, guilt-ridden and driven to align itself – particularly in regard to immigration – with the position of the Left in the attempt to become *acceptable*. The political illegitimacy of all forces except for the Republican Front resembles an implicit call for the return to a *single party* system, the mark of totalitarian regimes since 1793. In this *de facto* single party, the Republican Front, only *tendencies* are acceptable (as they were in the ruling Communist parties of central Europe). While these tendencies may 'democratically' alternate, the alternation of Left and Right is only apparent and cannot challenge the overall political line of the single party, which is Left wing.

The RepublicanFront, like the single party in the totalitarian former USSR, clearly no longer pursues any revolutionary aims; rather, it serves to consolidate existing tendencies in society. This temptation of the '*de facto* single party', concealed under the guise of a multiple party system, strongly emerged with the suggested ban of the Front National and the lawsuits brought to make Le Pen ineligible. It is one thing to wish to ban a small group, quite another to do the same with a party that gets 15% of the votes...

The system, which is running out of steam, is actually trying to operate a *democtomy*: a 'restrictive amputation of democracy'. This is where it's gotten. The same logic underlies 'representative trade unions', even if these are only a minority phenomenon. From Robespierre and the Soviet Union to the Republican Front, it is always the same process, albeit in a *soft* version today: people do get to vote – it's a democracy, after all – but can only vote for *acceptable* candidates –those of the party.

Embarrassed, in order to justify its anti-democratic policy the system always turns to its favourite obsession: Hitler, the big bad wolf. The argument goes like this: 'Watch out! Hitler came into power *democratically*' – the subtext being: we should limit, isolate and keep watch over this dangerous democracy and exclude all *unacceptable* parties. Now, this rumour stubbornly upheld by the Left does not withstand historical scrutiny: for just like Mussolini, Hitler actually came to power through a *coup d'état* – clearly, one not presented as such at the time.

Another view that was voiced at the aforementioned demonstration in Lyon was, 'The Front National is unconstitutional!' – another example of Stalinist logic.

A slogan shouted everywhere against the Front National was 'Against intolerance and hate!' Now, the very system that funded *Mathieu Kassovitz's (worthless) film La Haine (Hate),* an apology for the hate of ethnic gangs against the French, was here accusing a political party wishing to limit the violence wrought by these gangs of being 'hateful'.

The system accuses the Front National of being intolerant because it wishes to ban it. Does the Front National call for the ban of any enemy parties in its platform? The system is charging the Front with the sin of 'advocating exclusion' while seeking to exclude millions of voters from the political arena. This may seem like a bad dream but it's not – it's something that is quite naturally taking place.

Totalitarian or pre-totalitarian regimes are not content with reversing the meaning of words, as Orwell described in *Nineteen Eighty-Four* or is shown in the film *L'Aveu*:[70] they accuse and condemn their enemies by charging them with their own shortcomings. This is a form of exorcism.

One final observation: at the end of the aforementioned demonstration against the Front National – on Sunday the 3rd of October 1998 – a 'multiracial' concert by Cheb Mami[71] was intended to take place. This was not staged 'because of incidents caused by groups of youngsters', as the press discreetly reported. Actually, these incidents consisted of riots caused by gangs of immigrants from the *banlieues* of Lyon, who attacked the demonstration that was intended to support them!

Ethnic gangs are undoubtedly the best campaigners for the Front National. The system is increasingly playing the part of a snake biting its own tail.

From the Discourse against Selection to that against Exclusion: An Absurdity of Egalitarianism, which Severs the Branch on which It Rests

A parallel can be drawn between the Leftist discourse against *selectivity* launched in May '68 and the present discourse of the Left, which is

---

70 *The Confession* is a 1970 French-Italian film based on a true story about a Czech Communist official who was arrested and then brainwashed into confessing to crimes that he didn't commit.
71 Cheb Mami, the stage name of Ahmed Khelifati Mohamed, is a popular Algerian raï singer. In 2009 he was imprisoned for drugging his girlfriend in an effort to force her to have an abortion.

IDEOLOGICALLY DISSIDENT STATEMENTS

centred on an opposition to *exclusion*. Ultimately, the same process is at play: in wishing to push its egalitarian principles to their very limit ('always more!'), the ruling ideology is ultimately clashing with common sense and plunging into social absurdity. It is paving the way for an inevitable clash: either it will turn back – at the cost of great lies and difficult manoeuvres – or it will be swept away by a form of socially functional inegalitarianism.

The rejection of school and university selectivity, which aimed at replacing *equality of results with equal opportunities*, by a heterotelic effect, brought less social justice. The results, thirty years after the introduction of this perverse principle ('orientation replacing selection'), are: a depreciation of diplomas, which contributes to unemployment; a flight of brains towards Anglo-Saxon universities; a general worsening of teaching and a growth in illiteracy; the end of school as a place of competition and education, and its partial transformation into an unbearable jungle; the creation of a two-tier school system: one private, qualified and selective for the rich, and one public and under-qualified for the poor. Paradoxically, *the egalitarian opposition to selectivity launched in May '68 is one of the causes of the present 'exclusion'*.

Hypocritically, trade unions and governments have not dared to apply their anti-selectivity principle to scientific matters: for no one wishes to be treated by incompetent doctors – nor will a space agency hire engineers unless they have been chosen through a strict selection process...

By contrast, worthless BAs and junk diplomas in 'psycho-sociologies' or 'aesthetics' are handed out like sweets or leaflets to rows of good-for-nothings who will queue up at social security offices to get underpaid jobs as switchboard operators, pizza boys, or waiters at McDonald's. This is the outcome of demagogy and egalitarian ideology, which rejects reality and ignores – and has been ignoring for a while – social mechanisms.

This hate of selectivity rests on an anthropological prejudice: the notion that human beings are all 'equally gifted' – as Alain de Benoist put it, that 'anything is as good as anything else'. Hence, nothing has any value anymore, and gifts – as well as excellence – do not exist. It is unacceptable now for human beings to differ in their intellectual capacities, creative skills, and even characters. This view corresponds to the rejection of life so aptly noted by Nietzsche. All ideas of hierarchy are banished; and rather than arranging natural hierarchies and inequalities according to justice, an attempt is made to forcefully impose inapplicable egalitarian principles. But as this is not possible, it becomes a destructive phenomenon: ultimately, wild hierarchies have been created that progressively erode social rights. It is capitalism, with its lack of scruples, that takes care of savagely operating the *selection* the state does not have the courage to implement.

The *anti-exclusion* doctrine always rests on the same rules. At first it asks us to fight against poverty, according to a praiseworthy sense of social justice. Very well. But now the very notion of exclusion has been twisted: what we are asked for is to prevent any form of *discrimination* between nationals and foreigners, including illegal immigrants. The same absurd logic underlies the opposition to selection: egalitarian ideology clashes with facts which – like the Front National – are *unacceptable*, as Louis Mermaz claims.

Does the refusal to legally expel illegal immigrants from Africa, China, Pakistan, etc. implicitly mean to acknowledge the fact that any Frenchman is free to illegally move to these places? For things should be so, according to the logic of reciprocity.

The present egalitarian policy goes against international law, which is based on the principle of *reciprocal discrimination*. Foreigners are given privileges that are denied to our fellow countrymen in other countries. Why then should laws require public officials to be French citizens? This too is a form of exclusion and discrimination! The right to vote for foreigners? And why not for French citizens living abroad too?

Why should the news that some illegal immigrants have lawfully been expelled and sent on a charter flight make the headlines – knowing full well that they will make their way back as soon as they are given the chance and that tens of thousands more enter the country each year – without ever mentioning the massive and hasty expulsions of immigrants that take place in African and Asian countries?

This *de facto* inability to expel illegal immigrants constitutes an official violation of the law – for elected governments are yielding to the pressure of minorities which have usurped their moral authority – and also contravenes the *nationality principle* at the basis of international law. This is yet another sign of the decline of democratic values and of the twisting of the notion of 'Republic' at the hands of those who claim to have invented it.

Egalitarian ideology actually developed an abstract definition of the 'nationality principle' (reciprocal discriminations and advantages among countries) when the problem of immigration did not yet exist. Today it is incapable of respecting this principle and is reverting to its old, catastrophic madness: universalism, the idea of a world without boundaries, without 'airlocks', nourished by the infantile romanticism of 'citizens of the world' who envisage a 'global government'. It does not realize that the planet can only be administered jointly on the basis of diverse and impermeable blocs – not by a jumble that will turn the world into a jungle.

Opposition to selectivity and opposition to exclusion: the failure of these will bring about a cataclysm that will elicit a return to archaic solutions.

The Imposed Revolution

Only when on the brink of disaster – when economic hedonism has come to an end – will the European peoples find the strength to react against what awaits them. No effective solutions can be expected prior to the unleashing of the catastrophe that will most likely take place. People's power to resist has been sapped by consumerism, comfort, and the countless 'commodities' of the society of the spectacle. People are weakened by the slack life they lead, by their boundless individualism, by the dreams promoted via television and advertising, and by their virtual experiences. This is what the anthropologist Arnold Gehlen has termed 'second-hand experiences' – socio-economic opium. Societies based on conspicuous consumption – as Thorstein Veblen[72] noted in the early Twentieth century – have undermined their own economic and social foundations. They have destroyed their own dreams of freedom, emancipation, equality, justice and prosperity by pushing them to their very limit, to the absurd, so that by a boomerang effect these societies are no longer capable of resisting financial crises, criminal organisations, and the social upheavals they have caused. This is an example of the *dialectical reversal* that Marx and Jules Monnerot have described.

These societies have caused a global anthropological weakening, whereby all the immune defences of humanity are collapsing. The cure can only be a radical and painful one. We are heading towards a

---

72 Veblen (1857-1929) was a prominent American economist and sociologist. He is best known for his 1899 book *The Theory of the Leisure Class*, in which he postulated that the emerging upper class of modern society was unique in that it contributed little toward the maintenance or advancement of civilisation.

revolution that will make the Russian one seem like a brawl in comparison.

Educational Principles (I)

Everyone is talking of 'the failure of the educational system' and of 'violence in schools', but these are only the fruits of a system that opposes selectivity and discipline in the name of utopias it wishes to preserve like dogmas. The reason why hundreds of thousands of young people cannot find a job – hence unemployment and crime – is that the current educational system serves not to educate (education: from Latin *e-ducere*, 'to lead out of a condition of ignorance and lack of culture'), but rather to perpetuate itself as a corporate and guarded administration, promoting as it does a form of schooling that is dogmatic and inefficient.

Here are a few common-sense suggestions:

School should no longer be compulsory above the age of 14.

It should teach the 'keys to knowledge' and rules of social conduct through discipline.

It should follow three principles: selection based on merit; reward; and punishment. It should also have a degree of solemnity to it.

For students over the age of 14, schools and universities should no longer be free except for those who lack financial resources but are found worthy of receiving a scholarship once they have passed a rigorous selection.

The last of the above suggestions is not unjust, in the Platonic sense, as a rich but incapable student will be less successful in a selective school than a poor but capable one. For this reason, a very rigid selection will have to be made based on merit and competence. As Pareto[73] has shown, the more rigorous a (rationally planned) selection in a social system, the greater the turnover in the elite, so that the rich will not be able to enjoy the income from their social standing for long. In the present regime inspired by the far Left, and which goes against selection, the poor have an increasingly under-qualified educational system at their disposal: the gifted poor cannot have any success, while the ungifted rich can.

---

[73] Vilfredo Pareto (1844-1923) was an Italian sociologist whose theories were highly influential upon Italian Fascism. His principal work is *The Mind and Society* (New York: Harcourt Brace, 1935).

These simple principles, which have nothing tyrannical about them, will never be applied in the present system, for it is at the end of its tether. They are intended for after the revolution.

Selection and discipline: these archaic but effective principles are the basis of true individual freedom – the social justice of the future.

Today, instead of striving to rebuild things, it might be better to leave the educational system to collapse completely, given its inability to accomplish its task and the state's utter lack of interest in the matter. The new state that will emerge in the post-catastrophic world may take things in hand again.

Educational Principles (II)

Anthropologist Arnold Gehlen explained that freedom is born from discipline: for 'breaking in' (*Zucht*), as he put it, creates new skills. An effective education, he argued, one that brings freedom, must rest by the very constitution of man on effort, discipline, stimulation, sanction, and reward.

George Steiner,[74] on the other hand, when discussing the ancestral principles behind the Jewish education he had received as a child and was in turn giving his own children, made the following un-PC comments in the pages of a mainstream weekly magazine, 'When confronted with all that is done today to avoid causing anxiety and neurosis in children, I say that on the contrary, neurosis means creation, and that it is what helps us become human. Believe me, in making everything easy for children, we make them fragile, not only from an educational point of view but – what is worse – emotionally.'

Today, children – the 'young' – are treated like small gods. When they get bad marks at school their parents don't punish them but 'correct' their teachers by smashing their faces. All punishment is deemed illegitimate. This deification of childhood and youth paradoxically seems to go hand in hand with a statistical increase in brutality against children and paedophilia. Societies that are growing old treat children and adolescents in ambiguous and pathological ways: with adulation, excessive love and boundless permissiveness, but also with perverse cruelty and sexual sadism. Healthy societies, by contrast, in dealing with the young adopt a strategy consistent with the goal of transmitting collective vales and allowing talent to flourish: training and protection, strictness, and respect.

---

74 George Steiner (1929-) is a prominent literary and cultural critic based in America.

No return can be made at present to these archaic principles, which have been forgotten thanks to the ignorant utopia of egalitarianism. The future, however, will take care of reasserting them.

Conservatism and Repetition: The Senile Ills of Modernity

Charles Champetier, the editor-in-chief of *Éléments*, once shared the following thoughts with me (and they would deserve a book in themselves): 'Mass-media society destroys the traditional structure of knowledge and intellectual and cultural innovation, replacing it with repetition.'

As already observed by Walter Benjamin[75] – an exile from Hitler's Germany who commented upon the totalitarian nature of television in the 1950s, when it had just been introduced in the United States – the audiovisual sphere, like the contemporary electrovisual one (Internet, CD-ROM, videogames, etc.), reproduces models and values without creating any new ones, according to a horizontal and strictly commercial logic. The same may be said of advertising: it repeats, follows, but never provides any innovation. Models of society are copied according to a conservative logic, particularly in the sphere of ideas and solutions. False innovations are 'created' – mock innovations. Ideas and artistic forms are simply moving in circles. Modernity consists of nothing but repetition, parroting, conservatism (of forms as well as values), and scholasticism, all under the guise of innovation and *trendiness*. A gap is growing between the ruling ideology, which repeats humanist dogmas, and the technological, scientific and demographic realities which follow in the mode of urgency. The situation is growing increasingly unstable and signals impending catastrophe.

Once 'metapolitics' – which is to say the political application of *new* political ideas – was hierarchically organised. An *avant-garde* would progressively impose its new ideas. Today, in the reign of dying modernity, avant-gardes of this kind no longer exist. Even trends, be they intellectual or fashion trends, are no longer easy to interpret. Everything

---

75 Walter Benjamin (1892-1940) was a German-Jewish Marxist intellectual who was part of the Frankfurt School. Faye is undoubtedly referencing Benjamin's famous essay, 'The Work of Art in the Age of Mechanical Reproduction', although he is mistaken as it doesn't deal with American television since it was written in 1935, long before the TV explosion of the 1950s. However, it is certainly relevant to the TV phenomenon, since Benjamin predicted that art in the machine age, divorced from the traditions of other forms of art, would be unable to communicate any meaning apart from the political.

functions horizontally, as if by reaction – nothing but repetitions. This is very evident in the field of music: forms and techniques change, but contents stammer on the same way. Even in the field of technology, innovation no longer serves to 'change life'. The Internet is changing people's lives far less than the electric lamp or telephone have done. These are all signs of a world that is entering a deadlock – a prelude to its end?

The PACS Gag: A Model for 'Facsimile Progressivism'

The conservative Right sees PACS as the product of a 'homosexual lobby' – the notorious 'pink mafia' – that wishes homosexual couples of both sexes to be granted the right to marry and adopt children. This is not at all what it's about. It has been a long time since the homosexual world needed ruses of this kind to impose itself. Besides, gay couples do not last long and very few wish to live together for long and adopt children. So there is no need to panic.

PACS are no 'war machines against families' or a 'means to destroy marriage'. Those who wish to marry will never be dissuaded by the existence of PACS. Things are far simpler. PACS are a gag: they're one of those devices and symbolic measures adopted by a system that's reached its end. As it's no longer capable of solving real problems, it seeks to shift people's attention towards progressive pseudo-reforms that will make absolutely no difference. PACS are yet another example of the false freedoms and worthless 'rights' accorded in the name of an emphatic individualism – a way of concealing an utter lack of any political projects.

On the other hand, PACS will increase the financial pressure on French society (with 6 billion Francs lost through tax exemptions at a time when fewer and fewer funds are being allotted for families). Any couple, whether male or female – be it even a fake couple – by submitting a simple declaration to the public administration can enjoy financial benefits, succession rights and housing rights that are all paid for by society. In such a way others are being charged with duties in exchange for nothing at all. In the Napoleonic code, so full of common sense, it is assumed – as something quite logic and natural – that married couples benefit from financial advantages because they will renew society by producing offspring. Napoleon himself stated that 'concubines take no notice of the law, and the law takes no interest in them'.

The Left came up with PACS not so much because it sought to win the favours of the homosexual lobby, but because it made the following reasoning: our 'progressivism' is moribund, and we're no longer able to pursue social justice in any concrete way; the struggle against unemployment and poverty is beyond us. The only solution for the Left, then, is hypocritical progressivism. Hence the idea of PACS, which like other pseudo-humanitarian measures – such as the regularisation of illegal immigrants – brings not an ounce of good to the people, but only increases the burden on everyone's shoulders. Through this legislative device the Left seeks to give the impression it is being loyal to its progressive vocation.

Another point: PACS also help the Left and 'Republican' Right to create mutual artificial disagreement when they ultimately agree on almost everything.

The whole PACS affair illustrates the spinelessness and impotence of governments in this declining democracy of ours. The process is always the same. Incapable of solving *concrete* problems, governments feed public opinion *abstract reforms* that are always justified as further acts of humanitarianism and tolerance. True ills are not treated: the patient is only given painkillers (in the form of stupefying audiovisual or electronic entertainment), while pretending to solve false problems. Increase birth rates? Halt the desertification of 60% of our land? Prevent the catastrophe expected to take place after 2010 because of the failure of social budgets? Reduce urban pollution? Restructure European institutions? Don't worry about these things! It's all too complicated. Empty symbols are preferable: close the Superphénix nuclear plant[76] or establish gender equality in political parties. A tentacular, socialistic and taxing state gets all the fatter the more it lacks power, authority, and efficacy. The political class is powerless (for it lacks human qualities and determination) and is only interested in electoral propaganda. It lives by the day, with no concern for what could happen even in the short term; it makes no forecasts, and risks doing nothing beyond mock reforms. Yet it would do well to worry about the future.

Rap and Techno

From a musical point of view, rap – like techno – is a very poor genre. It is not open to any renewal. Its range of harmonies is too small and its

---

76 The Superphénix plant operated from 1985 until 1997, when it was shut down due to popular opposition. It was subjected to a rocket attack by eco-terrorists while still under construction in 1982.

rhythms too repetitive. Its lyrics, written by talentless people with public funding, are worthless, plaintive and falsely violent. NTM is nothing but subsidised propaganda and gratuitous provocation:[77] an aping of the tough Black bands active in the Bronx in the 1970s, minus the musical talent, power and sincerity. Utter impostors. The same goes for all contemporary rappers. It is working for the moment, but won't last long. MC Solaar is a good writer of lyrics trapped in a musical deadlock.

As for techno, it is not music but percussion. This 'music' also won't last long. It is devoid of any content. Techno and rap, like hip-hop, will go the same way as twist and disco, because they do not belong to any aesthetic or musical current, but merely provide a social look – and looks are transient things.

Rock'n'roll, on the other hand, is eternal, for it can take various forms and rests on a range of harmonies. It has managed to survive and remain in fashion. What is now spreading across the world, though, are ethnic forms of music: Latin, Asian, Celtic, Greek, Arab, African, etc. – renewed forms of popular music.

The Screen of False Freedom

One of the paradoxes of our society is that it allows the spread, in a humanist, tolerant and soft guise, of tolerance towards social violence and the erosion of public liberties. Faced with growing crime, insecurity and economic uncertainty, with an increasingly interfering fiscal policy, the restriction of the right to express one's political views, a disturbing increase in juridical errors, and the electronic monitoring of the entire population, the system no longer contents itself with falsifying statistics or turning people's attention towards public debates of no interest.

The system has now adopted the strategy of false freedoms. This consists in granting civil society what are presented as 'new freedoms', which are actually of no concrete interest, but have the advantage of being featured on the media. Things such as PACS, the requirement for quotas of women in electoral office, the banning of hazing, the *de facto* impossibility of expelling illegal immigrants, appeals to the independence of the magistracy, and the representation of students on school boards are all pseudo-freedoms that only constitute an additional burden for people. In such a way, mock emancipation is used to conceal the encroaching limits placed on our freedoms.

---

77 NTM, a French hip-hop band, is well-known in France for its opposition to racism and class divisions.

Concrete freedoms are being replaced by abstract and virtual ones. The same mechanism has been running since the French Revolution.

'Positive Discrimination' is racist and sexist

Many American states have adopted programmes and laws based on 'affirmative action', i.e., 'positive discrimination'. The word itself carries ridiculous contradictions. And the same thing is currently happening in South Africa as well...

Affirmative action implies an unconsciously anti-egalitarian attitude. It calls for a definition of the 'races to be helped' – hence it's racism. Should we help Arabs and Koreans? A 'racial scale' is thus implicitly established, based on notions of superiority and inferiority, which actually derives from anti-racist ideology itself. In the United States many minority spokesmen have felt humiliated at being listed among those benefiting from 'positive discrimination'. A woman writer of African origins in France has recently petitioned for a fixed quota of Blacks to be introduced into television programming.[78]

To put it briefly, women, Blacks, etc., are all being likened to the congenitally handicapped and underdeveloped, as people we should pity and help with a (considerable) push. What humiliation! 'White males' must be penalised so that others may find a place in the sun: but doesn't this very idea entail that 'White males' are intrinsically superior? Hence, this alleged superman must by discriminated against by authority in order to make way for 'others'. The subtext here is that women and Blacks are eternal victims who by their very nature require help: weak creatures to be constantly protected from oppression.

Anti-racist, egalitarian and feminist ideology is biting its own tail. It reasserts racist or sexist ideas of inferiority while claiming to fight against them. If I were Black I would be furious at being treated like a virtual incompetent who always needs help!

On the other hand, in forcefully imposing a 50% quota of women among political candidates, egalitarian ideology is going against its principles of equality and harming the sacred 'cause of women'. If most candidates are men this is not because of any conscious decision to leave women out, but because there aren't enough women standing for

---

[78] This petition was filed in 1999 by Calixthe Beyala, originally from Cameroon, who was also President of the Collectif Egalité. The Collectif asked citizens to refuse to pay their TV licenses until a quota was established. Although no formal quota has ever been set for French television, there has been an increase in the visibility of Blacks since her complaint.

elections. By imposing an equal quota for women, they are actually imposing a number of candidates who are bound to be mediocre. Suffice to recall the case of Juppé, who, wishing to prove that she was 'trendy', had six female ministers appointed in her government who were soon dismissed for their incompetence. Why not impose a 50% quota for men in jobs of great social importance – such as the magistracy or high school teaching – where woman are in the majority? Why not establish a 50% quota of women among doctors and surgeons, most of whom are men, by setting up separate entrance exams? But there's a snag here: perhaps the egalitarian partisans of positive discrimination would not be too happy at being operated upon by female surgeons of dubious talent.

Let's go one step further: besides gender quotas, why not also adopt ethnic quotas to reflect the presence of each ethnic group in our multiracial society? In such a way, Air France would have to recruit its personnel through 'ethnic colleges' and employ X per cent of Black pilots, Y per cent of pilots of North African origin, and so on. But this of course will never happen, because there's a limit even to madness.

Positive discrimination, whose aims are anti-racism and anti-sexism, makes society increasingly sexist and racist. When egalitarianism seeks to stretch its principles to their very limits, according to an abstract logic, it ends up perverting them and making them absurd and contradictory. Are equal opportunities not yielding equal results? Then we must forcefully impose this equality into the results by destroying the very notion of equal opportunity that lies at the basis of egalitarian ideology... All this is happening because the latter dogmatically refuses to acknowledge the *inequality in skills* among different individuals and ethnic groups. 'Nature' does not share our views? Then let us change nature by decree, as is done with history! An ambitious plan that leads straight to catastrophe... Well, so much the better: to quote an Indian proverb, 'When you see your enemy dancing on a rooftop, let him do so and applaud his feat.'

The Return of Class Struggle: the Left on the Exploiters' Side

According to the classic mythology of the Marxist Left, class struggle opposes wage-earning workers against the managerial or parasitic bourgeoisie. Today, the real class struggle is between the *wage-earners in the protected sector* – who can almost certainly count on a lifelong career and benefit from great privileges and acquired advantages – and *the unemployed and those with precarious or risky jobs*, categories which are becoming increasingly common. The former live off the latter and can use strikes as a weapon. One kind of worker derives its security from the uncertainty of the other. The paradox here lies in the fact that the contemporary Left and its trade unions – particularly those connected to the public sector – *are defending the exploitive and secure economic class: that of the protected wage-earners*. Increasing privileges, an unwavering preservation of existing benefits (funded with taxes from the chaotic private sector), a reduction in working hours for employees in the public and semi-public sectors and in large business groups (the 35 hours scam), etc.

The strikes organised in the winter of 1995-1996[79] were not the expression of any form of social defence, but of a *corporative class struggle*. The wage-earners from the protected sector were asking for further funding and more sacrifices from the unprotected classes who are the ones really creating wealth.

So while the Front National was gaining consensus among the proletariat of the risk-taking sector, the new unprotected classes and those who personally face risks to produce wealth, the new electoral battles of the Left were being launched by the bourgeoisie of the protected sector, the one safe from unemployment, poverty and crime...

As for the Trotskyist Left, it is stuck defending 'sans-papiers'. Through its theory that illegal immigrants cannot be expelled it is objectively accomplice to these immigrants' exploitation of national workers, caused by the fact that the latter are financially burdened by the arrival of aliens constantly in need of 'help' who are free to create businesses on the black market, thus harming the rest of the economy.

The far Left and class struggle: some honest and intelligent people on the far Left are aware of what is not working and why, but are incapable of suggesting any alternative models. They acknowledge that the system is failing to offer any credible social and economic solutions, and that raw liberalism leads to economic horror. Yet, they do not dare suggest possible answers or plans for society; on the one hand, because

---

[79] In November of 1995, a series of general strikes, primarily by railway workers, took place in reaction to some of the policies of the new government of Alain Juppé, which was accused of attacking workers' and women's rights.

IDEOLOGICALLY DISSIDENT STATEMENTS

Marxist strategies have failed; on the other, because they are starting to think – without ever admitting it – that the true remedies are to be found not on the Left, but in what Zeev Sternhell[80] terms the 'revolutionary Right' and Pierre Vial 'national populism'.

Actually, the Left has long abandoned the social sphere. Today, it seeks refuge in 'ethics' – a new fraud. It no longer cares about 'defending the oppressed', except in a pretend way; actually, it never did: the Marxist-Trotskyist tradition has always taken little notice of the 'working class' and 'proletarians' – and 'immigrants' today – whom it continues to treat as masses to be manipulated to stir up social chaos in the hope that its cynical and ambitious (as well as perfectly 'anti-Republican') circles may one day come into power for good. Unfortunately, it is not enough to merely seize power: power must also be preserved. With its pseudo-moral strategy, the Left and far Left have been playing with fire while forgetting the joker: Islam.

The Contradiction between Integration and Communitarianism

When considering the fate of immigrants and their offspring in France, both Left and Right fall flat on their faces. 'Republican' and 'humanist' principles lead to absurdly contradictory solutions in themselves: according to Republican logic, we are told again and again that *integration* is necessary, but at the same time that *assimilation* must be rejected, as this would be a form of racist coercion. Also drummed into us – usually by the same people – is the idea that we must preserve *differences*: this is the theory of differentialism or communitarianism, which believes in the harmonious coexistence of a 'Republican Islam' respectful of secular values alongside the virtues of communitarianism, i.e., of a viable and peaceful ethnic mosaic. At the same time, an apology is made for intermingling and race-mixing, which would seem to contradict the communitarian view whereby each ethnic group is to affirm its own identity... In other words, these people want to have their cake and eat it too; they want everything and its opposite: integration without assimilation, the preservation of ethno-communitarian differences and the melting pot, and so on. Once again, the ruling ideology can be seen

---

80 Zeev Sternhell (1935-) is an Israeli historian who specializes in the history of Fascism. He is particularly noted for having traced the roots of the ideology of Fascism back to French, rather than Italian or German, philosophy and politics. His works in English include *The Birth of Fascist Ideology* and *Neither Right nor Left: Fascist Ideology in France*.

to fall victim to its favourite vice: belief in miracles. Is banning chadors in schools Republican or racist? Or is it both? The intellectual acrobatics performed by the media and politicians in this matter show that they are caught in an utter deadlock. It should be acknowledged that in history, insurmountable contradictions exist, which is to say *insoluble problems*. Only a clear break can bring some solutions, but only through the painful establishment of a different system.

Vengeance, the Motor of Politics

Monte Cristo:[81] vengeance is the most accomplished form of political power. Just as in love, nothing gives us more energy than the desire for vengeance. It can last for centuries and will never disappear. Currently we are prey to the – perhaps unconscious – desire for vengeance of the peoples of the South whom we have colonised and who feel exploited and humiliated. Vengeance is one of the guiding forces of history. The First World War chiefly sprung from the thought, 'Give us back Alsace-Lorraine!' The fate of the Twentieth century was sealed when, following French defeat in 1870, Bismarck chose to re-annex these lands, which Louis XIV had conquered. [82]

A striking parallel can be drawn between emotional relations among individuals, political relations, and those among peoples.

The answer to all this cannot be 'we're wrong and you're right: we are waiting to be assaulted – punish us, invade us', as the dominant ideology proclaims; but neither can it consist of a hate campaign. The solution is to defend oneself 'with a detached spirit', to quote Demosthenes.[83]

Multiracial Society, Multiracist Society

At the cost of sounding repetitive, I wish to stress an important point again.

Recently, a news report on *Libération*[84] made the following distressing observation: in Brazil, a multiracial country with the most

---

81 Alexandre Dumas' famous novel *The Count of Monte Cristo* (1844) is about a man who spends many years and amasses a fortune in order to have revenge against those who had had him falsely imprisoned.

82 Alsace-Lorraine was a territory created by Bismarck after the seizure of territory from France along the German border following France's defeat in the Franco-Prussian War. This led to a great deal of resentment of Germany by the French, and was one of the factors leading to the First World War.

83 Demosthenes (384-322 BC) was an Athenian statesman and orator.

84 *Libération* is a Left-wing newspaper that was founded by Jean-Paul Sartre in 1973.

IDEOLOGICALLY DISSIDENT STATEMENTS

anti-racist Constitution of all, an impressive ethnic hierarchy exists and Blacks – with the exception of football stars (modern gladiators) – are considered the lowest of the low. Economic misery and social contempt: a sizable portion of the population is alienated through poverty, ignorance and crime. The brave journalist explained that, at the end of the day, apartheid South Africa was 'less racist' than anti-racist Brazil!

I know the United States well: with minor differences, the situation in this country is not far off from that of Brazil. Yet the article in *Libération* does not derive any practical conclusion from its observations, stuck as it is in its multiracial dogma. Its author believes in miracles and clings to his utopia, imagining that the situation might improve through 'education', 'tolerance' and 'good will'. It's always the same story: 'If all the children in the world were to hold hands',[85] as the song goes. The Leftist myth of *education* and *prevention*.

Egalitarian ideology has always despised the sociology of reality and human society as it is, was and will be. It imagines that the 'spirit of the Laws' knows no limits, and that decrees create reality. This dangerously naive outlook makes people stubbornly believe that a multiracial society organised according to anti-racist *laws* will be a harmonious one.

This is the worst of all egalitarian utopias. In history, ethnically heterogeneous societies have always been powder kegs. 'Non-racism' and ethnic respect can only exist among peoples living in separate systems as political entities. We have failed to learn this from the tragedy of Yugoslavia. Not *one* example in history exists of a multiethnic society that is not conflict-ridden, bitterly hierarchical and oppressive. But this lesson is simply ignored and dogmas prevail over experience.

Egalitarianism (just like 'communitarianism') imagines that ethnic differences can be confined to the private sphere by people coming together in the public, social and political sphere. This mechanistic belief has never been illustrated in practice.

In 1996, I met an avowed American racist – a wealthy ranch owner and J.R. type[86] – in Texas. He openly told me, 'I do not understand why certain parties are seeking to curb immigration in Europe. All these immigrants will serve as a new class of slaves for you! All you need is

---

85 This is from the song 'La Rèvolution'.
86 J.R. Ewing, a (fictitious) wealthy Texas oil baron depicted in the long-running American soap opera, *Dallas*.

an effective police force, like ours, to repress any riots.' What can I say? Many racists dream of a multiracial society...

In the United States, a vast territory and a country of immigrants, multiracial society brings only limited conflicts. It will be different in Europe, where space is more limited and Islam is a growing presence. We are moving towards many years of ethnic wars with no certain outcome. After all, these wars have already begun...

The Need for Revolutionary Thought. How to Define It?

The system is *globally* dysfunctional. No improvement is possible, for the ruling ideology – and not public opinion alone – is opposed to the idea. Incompatibility has emerged between this ideology and the practical solutions that would have to be adopted to save what might be saved. Today no partial and specific reform will suffice: the entire system must be changed, like an old motor which must be replaced, as its individual components cannot be repaired.

Any political party whose goal is not simply the career advancement of its members, but the salvation of its country, must cease thinking in reformist terms and embrace a revolutionary perspective. This can be described as a *state of war*. The 'classic' form of political opposition is where parties regard the power they wish to conquer as that held by their *opponents* and political *colleagues*; revolutionary opposition, by contrast, sees those in power as its *enemies*.

Two versions of revolutionary thought exist, as Lenin – following Machiavelli – had perfectly grasped. The first is the *siege* approach, which leads to failure. It is the strategy of the lion which ends up dying a brave death, pierced by lances. This strategy rejects all tactical alliances and temporary compromises in the name of a misleading notion of doctrinal purity. One sees oneself here as being under siege rather than as a conqueror. He leads the assault with gaudy red trousers, his moustache in the wind, only to be hacked down by enemy machine-guns.

The second revolutionary approach is *attack*. The means used here are subordinate to one's end. This is the strategy of the fox which always manages to steal the hens at night. Those who adopt it are willing to sign alliances with useful idiots and turncoats, and know how to hide a sword under their toga to strike all the harder. They know how to lay

IDEOLOGICALLY DISSIDENT STATEMENTS

ambushes and show patience and steadfastness, and to conceal their radical aims. They know how to make temporary concessions without forgetting about their genuine objectives, sustained by an iron will. They practice the art of deception which Nietzsche commended. Like good sailors, they know how to steer clear of obstacles and sail against the wind without losing sight of the harbour, their final destination.

The former revolutionary perspective is *Romantic*: it stems from our Germanic and Celtic roots. The latter is *Classical*: it stems from our Greek and Roman roots. The former perspective cannot lead to the seizing of power; but once power has been seized, it will find its rightful place once more.

The True Reasons for the Demonisation of the Front National

There are many points in the platform of the Front National with which I disagree – particularly its European strategy, economic doctrine, and latent Jacobin nationalism. Still, as Baudrillard wrote – and this caused him to be brutally ostracised by the intellectual class and the media – the Front National is the only genuinely revolutionary party to have emerged after the Second World War. Its clear aim is to overthrow the system. We can always disagree on the tactics to be employed or on specific doctrinal points, but what matters is to share the same global view of things. Despite all its defects, tactical mistakes, internal quarrels, and ideological blunders and inconsistencies, the Front National has become something that *cannot be ignored*.

Why is it being criminalised by the intellectual class, the media and the self-righteous bourgeoisie? Is it because it is 'racist', 'fascist', 'of the far Right' and 'anti-Republican'? Not at all. These accusations made by scared false virgins are only pretexts. No traces of racist, fascist doctrines are to be found in the political platform of the Front National; and on the other hand, its most embittered persecutors – including Jospin[87] and 50% of his Socialist ministers – belong to currents of thought that flirted with totalitarian Communism.

The true reasons behind the ostracism of the Front National are to be sought elsewhere. The Front prevents politics from going round in circles: first, by unmasking them and refusing to apply them, it breaks

---

87 Lionel Jospin (1937-) was the Prime Minister from 1997 until 2002, of the Socialist Party. He was a member of a Trotskyist group in the 1960s.

the rules set by the political establishment, i.e., those of careerism, based on the pseudo-Republican pact between Left and Right, consisting of false contrasts and real complicity; second, it engages in politics where it had been agreed that one should only engage in business; third, it has ideas and elicits debate where the general consensus is that ideas are dangerous (for they create divisions and stir people's conscience) and that the system based on widespread stupefaction of the masses at the hands of the elite of the society of the spectacle should not be called into question; four, it demands that the ruling power provide concrete solutions to practical problems, where it is evident that governments must 'communicate' and manoeuvre in order to be re-elected rather than 'attain success to win trust'; and fifth, it breaks the law of silence and tells the king he's naked by revealing catastrophic social and political truths.

To put it briefly, the Front National is being demonised not for any hypocritical moral reasons, but because it is *too democratic and too political*: *because it poses a direct threat to the careers of influential politicians* in institutional parties and various lobbies. It represents an enduring peril because it seeks to 'win people's trust'.

The Front National is not being demonised and fought against – often with hatred bordering on illegality – because it 'threatens the Republic'; but rather because it threatens the pseudo-Republicans. It is not attacked because it promotes unacceptable values, but because it *has* some values, and this in itself is unacceptable.

While I do not agree with many specific points of its platform, I must acknowledge that the Front National represents the first political force in Europe to implicitly embody an idea that is deadly for the system: passing *from resistance to revolution*.

The false elites that have usurped the Republic are trying to gag and undermine the Front because it seeks to re-establish the moral contract between the people and its leaders. Hence, it is accused of being immoral. But facts will speak for themselves – the politicians and the media will not be able to twist them. So the only path open to the system is not to ban the Front National but abolish the people. It is already trying to do so. Immigration is one of its weapons, but it is a double-edged sword, for the system – and I will stress this once more – is forgetting about an essential player: Islam.

Machiavellian Principles for the Seizing of Power

Let us read Machiavelli again, whose works Lenin and Napoleon knew well. Public opinion changes: the people of today would hardly put up with the kind of solutions and therapies that could cure its illnesses. Today the very railway workers who are being attacked by ethnic gangs would be happy to join a demonstration against the expelling of illegal immigrants! The fickleness of *quiet, cool times*... But if things were to turn *hot*, in times of serious crisis, all this would be different. When people have their backs against the wall and are suffering piercing pains, they easily change their opinions. Any revolutionary party must realise that it will only seize power if a crisis or emergency occurs that will make people accept what is currently unacceptable. This will never happen in the lukewarm climate of a situation that is slowly rotting, where propaganda is capable of neutralising any revolt or the stirring of the public conscience.

A revolutionary party must present itself as a saviour. Should there be an upheaval, the ruling ideology would disappear along with its taboos, and this would be the right moment to stand in the gap it has created. Revolutionary parties should envisage their action as following from times of crisis and chaos. To be revolutionary is to reason like a surgeon, not a reformer. Reformers will prescribe painkillers or break the thermometer. Revolutionaries will recommend a surgical operation and a treatment to eradicate the illness once and for all. Revolutionaries don't try to reform utterly diseased organic systems: they will change the whole regime, or – rather – *transform* it.

A revolutionary party should simply serve as a machine to seize power and exercise it as any other party would do. First of all, it should expect the first months of its rule to be stormy ones marked by much conflict. Hence, it should mentally prepare itself not to give in and be ready to shatter old principles, particularly as these will have been considerably weakened by the crisis and the emergency situation. Secondly, a revolutionary party that has come into power must *create irreversible situations*, which no loss of power could threaten to abolish. It must strike fast and hard – and this will be accepted, as the rules of the game will have changed. The old values and taboos will have crumbled. The party will have to play off the fear it arouses. Finally, even in this media age, it must *put practical results before symbolic measures*. The man of the street must concretely perceive the effects of the new political programme on his everyday life. The qualities required here are imagination and perseverance.

The danger any revolutionary power faces is to believe that the old rules of the game are still valid. Actually, everything will have changed in the aftermath of the chaos. It is often said that any such power will have to face 'isolation on an international level'. But why should the international scenario itself remain unchanged? And besides, any precautions to be taken – as in the old world – will be of little importance compared to the crucial imperative of implementing the revolutionary plan. To quote Machiavelli, 'The new prince must be determined and courageous above all.'[88]

The Left is neither Reformist, nor Revolutionary or Conservative: It is a Means of Reinforcing the System

Something evident that we should always bear in mind is the fact that, since the mid-Twentieth century, the Left has been feeding off the myth of revolution and reform. It passes itself off as being against the system, when it *is* the system. It passes itself off as being oppressed, when it oppresses.

The reforms promoted by the socialist Left, which reinforce the status quo, merely serve to further strengthen the influence of its own ideology. As for the far Left, which currently seems to be undergoing a renaissance, its role (like that of the Greens and the Communist Party) – now that the project of establishing a Communist society appears ridiculous – is but a more pronounced form of the socialist Left: to reinforce the ideology and structures of the egalitarian machine, particularly in its favourite field: immigration. The far Left serves to accentuate, accelerate and absolutise the trends in contemporary society, turning them into something definitive.

It is no longer a matter – as it was in May '68 – of 'changing society', but of pushing egalitarian society to its very limits. The far Left has given up on the idea of drawing up plans for a different society. It no longer engages in anti-capitalist and anti-bourgeois tirades; it doesn't even have enough power of imagination to develop a new version of Communism (as the Frankfurt School[89] attempted to do). Its discourse

---

88 From Chapter 18 of *The Prince*. See chapter 1, note 27.
89 The Frankfurt School refers to a group of Marxist philosophers who opposed Communism, capitalism and Fascism. It remains highly influential in the area of cultural theory. Some of its most prominent writers were Walter Benjamin, Theodor Adorno, Herbert Marcuse and Erich Fromm.

IDEOLOGICALLY DISSIDENT STATEMENTS

is limited to the same old lament: 'Let us proceed further along the path of egalitarianism!'

While criticising 'exclusion', it fails to suggest any alternative social or economic model. It has obsessively re-centred its doctrinal line on a moral question: helping immigrants – who are falsely regarded as the only outcasts – and the promotion of de-Europeanisation on an ethnic and cultural level.

The reforms promoted by the Left are mock reforms: nothing is reformed and nothing solved; what exists is simply reinforced – particularly our present crisis.

The Great Imposture of the Greens, the Kings of Concealment[90]

In France as well as Germany, what is paradoxical about environmentalist politicians is that they engage in neither politics nor environmentalism. The political platforms of the Greens contain no *real* environmentalist suggestions, such as the transport of lorries by train instead of on highways, the creation of non-polluting cars (electric cars, LPG,[91] etc.), or the fight against urban sprawl into natural habitats, liquid manure leaks, ground water contamination, the depletion of European fish stocks, chemical food additives, the overuse of insecticides and pesticides, etc. Each time I have tried to bring these specific and concrete issue up with a representative of the Greens, I got the impression that he was not really interested in them or that he had never really studied them.

Brice Lalonde[92] once discretely informed me that the true target of the Greens is nuclear energy, which they demonise as a sort of magic force and associate with the atomic bomb. Now, the Greens' explicit goal of closing down all nuclear plants entails the reopening of all oil and carbon plants, which are far more polluting and dangerous (not to mention expensive). The fight against nuclear plants thus goes against environmentalism. The Greens are voicing few protests against the black sea of petrol and the carbon dioxide emissions by which we are

---

90 The French term Faye uses is *'les rois du cache-sexe'*, which literally translated means 'kings of the G-string'. He uses this term on several occasions to describe politicians who try to conceal the truth.

91 Liquefied petroleum gas, or autogas. It is an alternative fuel which is low in carbon emissions.

92 Brice Lalonde (1946-) is a leader of France's Green Party, and has run for President in several elections. In recent years he has become known for his Right-wing positions and associations.

engulfed, but go off as soon as the slightest nuclear incident occurs. The fact is, the Greens don't dare take on the international petrol lobby, which are no doubt happy to cough up some dough to intensify the struggle against nuclear energy. The national nuclear lobby is a far easier enemy to face.

All energy sources are polluting to some extent, and at the moment nuclear energy is the least dirty among those that can serve an industrial purpose. It is extraordinary to think that, in order to replace the least polluting energy source of all, the Greens are willing (as in Sweden) to make further use of fossil fuels – the most polluting energy source. The five alternative and less dirty energy sources currently available (geothermal, solar, wind, tidal and hydraulic) are technically incapable of providing the number of megawatts required by an industrial country.

Like the far Left in the economic and social domain, the Greens are happy to simply criticise and demolish. No study or serious suggestion has ever come from their ranks as to how to improve the use of the aforementioned energy sources – which are extremely clean – or come up with new ones. Possible suggestions would include decentralising the production of electricity by installing underwater dynamos in all rivers – a contemporary version of watermills – or set up windmills along windy shores, a plan for which has been drawn by a Dutch-Flemish company.

The concrete measures the Greens have taken, once in power, are truly laughable. It is enough to consider that Mrs. Voynet[93] has managed to block the construction of a channel between the Rhine and Rhône, causing an increase in the traffic of goods via lorries between the North Sea and the Mediterranean, which will become even more chaotic, expensive and polluting.

The Greens actually couldn't care less about environmentalism, which serves as a mere pretext for them. Proof of this is the fact that in Germany and France they go out of their way to defend the naturalisation of illegal immigrants, prevent them from being legally expelled, and so on, while doing very little indeed to serve the environmentalist cause. Environmentalism is only a mask for Leftism.

---

93 Dominique Voynet (1958-) is a member of the Green Party and was Minister of the Environment between 1997 and 2001, known for her environmentalism and pacifist stances. She is currently a Senator.

Political environmentalism, as shown by Greenpeace campaigns, is a large-scale fraud. Like many charitable, humanitarian and cultural associations, it is only one of the countless disguises the political far Left uses to move its pawns and compensate for its notable lack of any alternative socio-economic project.

The Real Causes of Immigrationism: Xenophilia, Ethno-masochism and Electioneering

Why do all Leftists favour immigration? Why is it that the more people are to the Left, the more they welcome unrestrained immigration? The reasons invoked are both sophistic and ridiculous:

First, the needy and refugees must be let in to uphold the honour of France, an open society, and the place where human rights were first formulated. According to this view, being a patriot means making one's fellow countrymen support aliens who benefit from public aid more than they themselves. Being a patriot thus means transforming the anthropological, ethnic and cultural substratum of one's country within one generation – an unprecedented phenomenon in French history. The second reason invoked is that, because of their birth rates, native Frenchmen are no longer able to provide a generation turnover; hence, immigrants are needed. This is a magnificent sophism indeed: why not simply take measures to increase the birth rates of the native French? Well, because *natalism* is seen as a political and ideological sin. So let us turn now to the real reasons behind immigrationism. The first is a psycho-ideological reason, while the second consists of a political plan.

First reason: the Left, which spearheads immigrationism and is followed in this by a guilt-ridden Right, suffers both ideologically and morally from a sort of binary complex: *xenophilia* and *ethno-masochism* – the idealisation of African and Asian foreigners, and hatred of its own roots. This is reminiscent of the syndromes affecting the anti-bourgeois Marxist bourgeois, anti-clerical defrocked priests, and anti-Semitic Jews. If applied to Leftist ideologues, political psychoanalysis would reveal that these people regard the 'White man' as being intrinsically guilty and stained by the unforgettable and unpardonable sin of having exploited non-European peoples (through colonialism, racism, etc.). Immigrationism and theories promoting the idea of a multiracial, mixed society thus represent *the work of being redeemed for our sins*. We must make up for our faults by disappearing as a homogeneous folk and allowing ourselves to be colonised and dominated. (By 'us' here I don't mean Leftist ideologues, but the hateful native European masses). One example for all: for work reasons, I often visit the world of showbiz. In the course of an interview with the beautiful and talented actress Béatrice Dalle,[94] who has a perfectly Leftist outlook and a pseudo-rebellious style, I asked her, 'Why don't you have any children?' She answered, 'They would make me fat. But I love children and would be happy to adopt some, if possible.' I then asked, 'You mean you would like one of those Rumanian or Ukrainian orphans?' Her answer – no further comment needed – was, 'No. I don't want to adopt any European children. Only coloured ones, from Africa or Asia.' What an interesting psychoanalytical case this would be: might it be that the ethno-masochism and xenophilia of the Left stems from a *racial obsession*?

The second reason behind immigrationism has to do simply with political and demographic plans. According to official statistics on naturalisation, birthright citizenship[95] and lax immigration laws, the number of voters of immigrant origin is constantly growing. Now, the vast majority of these people will vote for socialist parties and the far Left, which they regard as their 'protectors', while the native French working class – the traditional reservoir of votes for the Left – will turn their backs on them and choose the Front National. The plan here is a very simple one: to increase the population of immigrant voters; and then, to make it easier for them to vote by automatically enrolling them on electoral lists (this used to be a voluntary, 'civil' process). This is a short-term plan, but one that effectively serves the career interests of politicians of the Left and far Left, i.e., to ensure a lasting majority of votes to preserve their own power. For demographic reasons, Right-wing voters will be a minority for a long time. If our folk is not good enough – so the reasoning goes – then let's replace it with another.

---

94 Dalle (1964-) is best-known in the English-speaking world for her roles as a taxi driver in Jim Jarmusch's *Night on Earth* (1991) and as a cannibal in *Trouble Every Day* (2001). She is also known for her run-ins with the police.

95 Birthright citizenship, or *jus soli* (law of ground), is in opposition to *jus sanguinis*, or the right of blood. Most European countries have a policy of blood citizenship in which one's eligibility depends at least partially on one's ethnicity, although France and the U.S. have a policy of birthright citizenship, by which citizenship is available to anyone born within its territory.

National Preference: A Self-contradictory Notion

The debate on 'national preference' is not unlike debates about the Loch Ness monster: it concerns something which fades quickly from view. The so-called Republican Left and Right see national preference as a fascistic and discriminatory idea. Those municipalities which provide subsidies for couples of French origin are seen as breaking the law, like all charitable associations that limit their help to French citizens only. Yet, according to the French Constitution, national preference is what regulates employment in the public administration, whether civil or military. So the Constitution itself must be fascistic and discriminatory: why not reform it immediately?

The whole of international law is founded on the notion of national preference. It is applied by all countries of the world, which systematically put their own citizens first, particularly when it comes to jobs. So all countries in the world except France must be fascist – as must be the national preference laws the parliament of the Popular Front[96] passed under Léon Blum![97]

Actually, both those opposing the idea of national preference and those supporting it are victims of a *self-contradictory political concept*. Egalitarian ideology simultaneously embraces the idea of *nation* and *non-discrimination*, of belonging and of non-exclusion. In order to consistently pursue the path of individualism and universalism to the very end, the ruling ideology must ultimately sacrifice the concepts of nation and *citizenship* so dear to it. We are all 'citizens of the world' – that goal is clear – yet not of any specific country. The very idea of the nation, like national citizenship, is now meaningless. As is, to some extent, the 'state'.

Do the Left and far Left, these great enemies of national preference, not realise that they are severing their very link to the nation-state and threatening their own doctrines regarding economic state intervention? Do they not realise they are implicitly siding with ultra-liberalism, which is based on the belief that there are no citizens but only individual atoms, disembodied economic subjects with no roots? The

---

96 The Popular Front was a coalition of Left-wing parties in the 1930s. It held power for a single year under Blum, in 1936-37. Faye is referring to the Blum-Viollette proposal of 1936 for Algeria (which was still a French colony), by which only the minority of educated Algerians would have become eligible for French citizenship. The proposal was never put to a vote because of massive opposition from French colonials in Algeria.

97 André Léon Blum (1872-1950) was the first Jewish Prime Minister of France, and held the office three times. He was imprisoned at Buchenwald by the Nazis.

most stupid Left in the world, against all evidence, is ignoring the fact that the rejection of national preference is the central dogma of ultra-liberalism. It has never read Milton Friedman.[98]

This demonisation of national preference is actually a residue of the Marxist idea of proletarian internationalism, which had been abandoned by the builders of Communism early on because of its utopian character.

The entire debate on national preference is a case of *the emergence of repressed notions*. It is a matter for political psychoanalysis. It is the Front National that first brought attention to the issue by formulating the debate on a semantic level. The Front has made explicit a concept that is implicit in Republican ideology, forcing 'Republicans' to recognise that it is incompatible with its own egalitarian and individualist dogmas. The self-righteous advocates of political correctness find themselves caught in an ideological trap: fighting the idea of national preference while emphatically defending 'citizenship' (or embracing 'French' patriotism and the idea of France) will prove an increasingly difficult acrobatic feat. On the other hand, the Left is being forced to confess its hidden thought: that a Senegalese enjoys all the rights of France, but a Frenchman enjoys none in Senegal. This disregard for common sense can't go on for long.

In drawing attention to the issue of national preference, the Front National has not managed to escape some of its own inconsistencies: for thanks to naturalisation laws, demographic trends and immigration, those it considers 'foreigners' are legally French by now, and this applies to the majority of young North Africans and Blacks.

Ethnic Preference: An Archeofuturist Notion

North Africans and Blacks in France who are legally 'French' have spontaneously ceased to reason in terms of nationality. They are Archeofuturist without knowing it: for they employ *ethnic* terms. They speak of the 'Gauls', 'white cheeses' and 'sons of Clovis'[99] to refer to French natives. What a gap between the official ideology of egalitarian naturalisation and social reality...

The dilemma the Front is facing is that its 'national preference' imperative also applies to the majority of young people of immigrant

---

98 Milton Friedman (1912-2006), an American economist, is regarded as one of the greatest economists of the Twentieth century. He was a staunch defender of monetarism and the free market. Perhaps his most important work is his 1962 book, *Capitalism and Freedom*.

99 Clovis I (466-511) was the first King to unite the Frankish tribes. He also converted to Catholicism.

stock, and this poses a serious problem. It is very difficult to argue that the notion of 'French nationality' is simply disappearing.

What would the solution be? The ruling ideology and its system are plagued by contradictions, which are bombs waiting to go off. It is the resulting *clash* that will solve the problem, not the ideologues of the system. It will then be necessary to make things clear and choose either to abandon the idea of *nationhood* completely in favour of a global individualistic and cosmopolitan outlook – the logical outcome of all egalitarian ideology stemming from Judeo-Christianity and the French Revolution; or to clearly adopt the principle of *ethnic preference*. This would be based not on an individual's formal and legal belonging to a given nation-state, but rather on his belonging to an ethno-cultural community. At the moment we are sailing in the fog through compromises and cheats. But I'm sure that events not far away will make things much clearer.

One last point: the etymology of the word 'nation' has been completely erased by the Left. The Latin root of this term means, 'a group of people born of the same stock' – in Greek, *ethnos*.

The Revolutionary Principles of Enmity and Friendship: A Critique of Carl Schmitt (I)

Carl Schmitt's central idea is that the essence of politics consists in *identifying one's enemies* and not in the liberal idea of an *arbitrary and peaceful administration of the state*. He is only half right. As some of his detractors have noted – people who shared his opposition to liberalism – the essence of politics also consists in *identifying one's friends* – the comrades who share our struggle. The Marxists had understood this well, without ever managing or daring to state it. Rather, they had given a utopian and mistaken formulation to the notion of 'comrade', which they limited to 'class comrade'. But this is a false, abstract idea with no anthropological basis, just like the concept of 'citizen' that was formulated during the French Revolution.

It is reasonable to assume that a political force, party or movement will not succeed in its goals unless divergences among its members – whether sincerely ideological in nature or simply motivated by personal ambitions – are weaker than their drive to fight their common enemy. Still, external enmity is not enough in itself to consolidate a party: internal, disinterested friendships and shared views must also exist within its ranks.

It is not enough to fight against a common enemy. A genuine community of values must also exist, based on purely positive *feelings*. A comrade is not simply one's ally in a struggle. Without comradeship, any cunning enemy can divide a party.

Internal friendship must be as strong as external enmity. People can hate the same enemy without this healthy aversion of theirs lessening their mutual enmity. Lenin wrote, 'Let us unite now – we shall settle the score later.' What he meant by 'later' was 'once we have seized power'.

A subtle dialectic exists between friendship and enmity. A political movement can hope for success if its internal disputes never break out, for underlying friendship among its members prevents their mutual disagreements from turning into public, open conflicts. Trotskyists and Leninists waited until they had seized power before – tragically – parting ways under Stalin, the heir of the Leninist current of 'Russo-Bolshevism'.

Internal enmities must always disappear in the face of external ones. In other words, the unity of a political movement cannot be based exclusively on external enmity, as Carl Schmitt suggests. This is a mechanistic view of things. A party can only find its unity in the mutual friendship of its members, in their sharing of common values that transcend any doctrinal or tactical disagreements between them.

Carl Schmitt is right in rejecting the liberal view of politics as the neutral 'administration' of the state. But in limiting the essence of politics to the *identification of one's enemy*, he only goes halfway and forgets an essential point. His definition of politics lacks a positive dimension, both spiritual and anthropological. The essence of politics also includes the *identification of one's folk and who is part of it*. It implies an answer to the question: *why are we fighting* – for what values? This is an affirmative view of politics: a constructive, organic and long-term view, not a merely critical and mechanistic one. Politics are not a football match: it's not simply about defeating an enemy team, it's about developing a positive project. Between liberalism, which confuses politics with administration, and Schmitt's school, which limits it to the identification of one's enemy, a third path exists that I will attempt to outline in the following section.

What is the Essence of Politics?
A Critique of Carl Schmitt (II)

Carl Schmitt's idea of 'identifying one's enemy' is a crucial one. It must certainly be integrated into the overall definition of politics, of which it constitutes the essence – the axis and foundation.

The essence of politics might be defined as the *formulation and accomplishment of the destiny of a people*. This implies hostility towards an enemy, but also a voluntaristic reflection on a project of civilisation. I feel that the Nietzschean concept of 'will to power' – understood as something pertaining to historical development and not mere warmongering – could help formulate the essence of politics.

Today we are witnessing the death of politics. Politicians are merely fighting for a semblance of power, where no concrete project exists. Political authorities have no real power not because of the influence of financial, economic mechanisms, but because they lack a will to shape the destiny of their people – they lack a historical vision. The last politician in France was de Gaulle.

The essence of politics – which epitomises the qualities each genuine head of state must possess – is *aesthetic* and *architectural* in nature: for it consists of a long-term vision of a collective future. The true politician is an artist, a drafter of projects, a *sculptor of history*. He is someone who can immediately answer the questions: who is part of my people and what are their values? Who are its enemies and how can we fight and defeat them? And finally: What destiny should we choose to acquire power and carve out a place for ourselves in history?

The essence of politics pertains to historical development. It consists in *building* a civilisation, starting from a folk.

Liberals, who confuse politics with administration, and Carl Schmitt, who limits it to the identifying of one's enemy, both reduce politics to economics, with its petty rules of management and competition.

The idea I have suggested for the essence of politics is an archaic one. Pharaohs were known as the 'architects of Egypt'. Mine is tomorrow's solution: Archeofuturism.

The Role of Sex in Ideological and Political Repression – What about Prostitution?

It is interesting to observe that the increase of taboos and proscriptions in the field of political and ideological expression go hand in hand with a collapse of sexual taboos. Pornography (virtual sex one does not personally engage in) serves as a safety valve. It is like a theatrical set – a *papier-mâché* facade. People are free to consume X-rated material in all types of media, provided they think correctly. 'Tits on the telly', but no deviant ideas. Censorship is letting harmless subjects go in order to focus on more crucial ones. You have the right to put your fingers in the cookie-jar, but not to criticise the regime.

That being said, any repression of pornography would be stupid. The hardest blow that could be dealt to the sex industry would be to make brothels legal again, with medical check-ups and compulsory condom use. Virtual sex would thereby be replaced with real sex.

Whether state-owned or a registered private one, it makes little difference. So here is another archaic idea: to reopen medically regulated brothels.

Organised, legal prostitution is the best known way to channel deviant sexual energies, and to curb pimping and all forms of crime connected to uncontrolled prostitution. All ancient civilisations have known this.

Women who sell their bodies are not to be despised – certainly, far more despicable are politicians who make a profit off of the fake love they parade for their country. A prostitute is a proletarian like any other: she sells her work to the highest bidder, but she doesn't sell her soul. Would it not be wiser to make legal again and regulate the oldest profession in the world? The state would become a pimp again, but this would always be better than being a dealer – for the state taxes alcohol, tobacco and petrol, all of which are clear causes of death. In organised and controlled brothels people run no real risks – not even of catching STDs.

For the time being, society cannot accept this solution, for it is puritanical in the very fibres of its permissiveness.

Misguided Theories about Drugs

Compared to alcohol and tobacco, or unnatural industrial food, drugs have a very limited impact on public health (each year in France 10,000 people commit suicide – far less than those killed in car accidents – but only 600 die from overdoses). The crucial issue is that on a global level drugs feed mafias which generate a considerable turnover. Thanks to corruption, these are capable of defying states all over

IDEOLOGICALLY DISSIDENT STATEMENTS

the world and of funding terrorist groups. Drugs also lead to uncontrollable crime in society itself. So *the drug problem is a political and social, not a medical one.*

Drugs also pose an embarrassing question for environmentalists, who are known to defend the use of soft drugs: in countries such as Morocco and Columbia, 60% of the forests have been destroyed to make way for cannabis crops.

The mass use of drugs among young people, which began in the 1960s, can be seen as a search for artificial paradises in a disenchanted world – a way of creating a semblance of communitarian warmth in a world without genuine, living communities. This is precisely the syndrome Zola refers to in *L'Assomoir*,[100] where he describes the Nineteenth-century working class finding refuge in absinthe.

People should stop pitying drug addicts as they do certain countries of the Third World plagued by civil war and poverty: junkies *are responsible for their own destiny* – let's at least give them their due. Enough with charitable do-goodism.

As for the question of whether I have ever taken any drugs myself, I must answer: yes, of course. I have tried them all, even the worst: VDA, a brew made from birch bark treated with acetyl-salicylic acid (extracted from greenbrier), the base ingredient for common aspirin and a substance Siberians have used since the mists of time. Down there, in the Verkhovyansk area, locals call it 'vodschkaia', which means 'super-vodka'. Compared to a 100 ml glass of this bluish liquid, a line of coke is pasteurised milk. *Vodschkaia* kills...

The system strives to make drugs chic, cool and trendy. After all, this has been going on since the aftermath of the First World War, when coke came into vogue in certain shady bourgeois milieus. This is all implicit. It is acceptable for bands to get loaded, and for the stars of the showbiz, jet-set society and politicians (who are all part of the same world) to keep on snorting until they've ruined their noses. Drug trafficking is allowed to prosper in areas beyond the law's reach just so that we can have some peace; then, from time to time, some extreme measure is taken. The message that is so cunningly being conveyed is that a person who has never taken any drugs is a fogey – a bit of a virgin.

---

100 The title is regarded as being untranslatable, although the term was used in the Nineteenth century to refer to shops that sold cheap liquor, deriving from the French expression 'to get hammered'. It has been published in translation under its original title as Émile Zola, *L'Assomoir* (Oxford: Oxford University Press, 1995).

With extraordinary subtlety in the media, the ruling ideology is striving both to promote the use of drugs – by openly showing tolerance towards people known to be heroin addicts, for instance – and to exercise a form of repression that is as ineffective as it is hypocritical.

Most people who talk about drugs – whether to denounce their use or to hypocritically defend 'soft drugs' – know little about the matter. They may have had a puff or two of bad weed, snorted half a line of coke cut with saccharine (and for which they paid five times its usual price), or swallowed a placebo fake Ecstasy pill at some pseudo-rave party. And in the end, it's the rum and Coca-cola that got them high...

Legalising soft drugs would bring the state a number of advantages: it would provide an additional tax revenue (as from tobacco and alcohol) to make up for its inexhaustible lack of funds, and would deliver a blow to cannabis and hash dealers, thus presumably curbing the crime connected to this traffic. Still, the wiseacres of the Right – from Pasqua[101] to Madelin – who foolishly wish to give the impression of being modern and try to appeal to the young by making similar suggestions, forget that legalising cannabis would mean that dealers would focus on hard drugs. We would thus have an increase in the consumption of both legal cannabis and hard, illegal substances, and a rise in crime as well because more money would be circulating (a gram of cocaine costs around 800 Francs, almost as much as plutonium).

It would indeed be profitable for the politicians of certain countries to witness an increase in the international trafficking of hard drugs: this is an important source of funding.

Another interesting factor which no one dares to bring up – particularly among journalists – is the fact that the media and political elites or pseudo-elites make massive use of drugs, and particularly cannabis and cocaine, both in France and in the United States. The strategy the system is adopting on a global level is thus an extremely hypocritical one: forms of repression are being organised that are intentionally ineffective. Big traffickers are never caught – 'extreme action' is only taken by making the occasional seizure or publicising the capturing of some small dealer, who is served on a platter to public opinion. Alternatively, broadcasts are shown of mock military

---

[101] Charles Pasqua (1927-) is a French Gaullist politician. In the 1995 election he broke with Jacques Chirac to support the RPR candidate for President. More recently, he was accused in Iraq of being one of the people who illegally profited from the Oil-for-Food Program during Iraq's trade embargo between 1995 and 2003, and in 2009 he was convicted of illegal arms sales to Angola, for which he was sentenced to prison for one year.

IDEOLOGICALLY DISSIDENT STATEMENTS

operations organised with the support of GIs in some poor country where illegal plants are grown.

On a planetary level, it is quite clear that there is a will to allow the lucrative drug business to thrive – and to *manage* it. The system has no intention of curbing the traffic in drugs, but only to *limit* it and profit from it; so much so, that new synthetic molecules are making their way onto the market that are cheaper, more effective and more specific in their effects than natural drugs of plant origin. This will be yet another problem to face...

The Theory of the Three Levels

In the *Dictionnaire idéologique*[102] I wrote over ten years ago, I distinguished three levels of political perception: first, 'worldview', a global perspective that entails an idea of civilisation as a goal and some general values; second, 'ideology', which consists of the explicit formulation of this worldview and its application to society; and third, 'doctrine', which simply concerns what tactics to use.

Skill for revolutionary movements lies in knowing how to act on these three levels.

The disputes between 'pagans' and 'Christian traditionalists' are a secondary matter, as are the contentions between those who romanticise France and those who romanticise Europe as a whole. What is essential for those with revolutionary ambitions is the first level: that of one's worldview. Secondary problems can be dealt with at a later stage.

Immigration and European Democracy

The increasingly felt presence of Islam in Europe and the growing influence of Afro-Asiatic cultural traditions on our continent – both of which are consequences of unrestrained immigration – represent a threat to democracy.

Out of altruism, 'they' imagine that education, reason and the 'Republican' spirit will prevail over the ancestral cultural traditions of immigrants. This is Régis Debray's[103] error of judgment. It rests on the myth of *spontaneous education* and *innate wisdom* embraced by the rational optimism of the Enlightenment. On the contrary, democratic

---

102 *Ideological Dictionary*. This is another title for Faye's book *Pourquoi nous combattons: manifeste de la résistance européenne* (Paris: Æncre, 2001).
103 Jules Régis Debray (1940-) is a prominent French Marxist intellectual. He is famous for having been a part of Che Guevara's ill-fated guerrilla insurgency in Bolivia in 1967.

virtues are ethno-cultural products limited to the European sphere: they are neither universal nor inborn in humans. By its very nature, democracy is something very fragile: its Greek founders soon lost it, as did the Roman Republic. Iceland is the only place that has historically preserved its democratic system for over 900 years. Democracy is menaced by social laxity and the media pretensions of *public opinion* – which consists not in the opinion of the public, but in that of active minorities – as well as by the rule of judges who seek to prevail upon the general will and correct its laws (as illustrated by the arrogance of the French Conseil Constitutionnel).[104] Even more simply, democracy is threatened by the establishment of a 'culture of everyday behaviour' implying one's submission to manipulation from sophisticated apparatuses.

It is indeed the case that a society may have ceased to be democratic and be incapable of securing the freedom and well-being of its citizens even if its institutions have remained formally democratic – it is enough for oppressive social practices to be repeated, tolerated and legitimised without necessarily having been made legal.

The culture of the 'young born from immigrants' so admired by the media is winning increasing social acceptance while affirming perfectly anti-democratic values. 'Black and North African culture' and the behaviour of its exponents, amplified by the propaganda delivered through rap music, promote attitudes and feelings that stand in open contrast to the apparent convictions of the politically correct elites that support them: chauvinism, gang culture, aggressive tribalism, a racial view of society, a ghetto spirit, contempt for women, the cult of gang leaders, a valorisation of brute force (the opposite of 'strength'), a rejection of all social responsibilities, an apology for group delinquency, an utter contempt of France and the 'nation', etc. The new 'ghetto culture' promotes social and political values among the young – and hence among future generations – that are the very opposite of those of the famous 'Republic'. To believe that through 'education' and 'persuasion' it will be possible to bestow a sense of 'civic responsibility' upon outlooks utterly shaped by the aforementioned values and forms of behaviour once again reveals that *belief in miracles* is the senile illness of Western ideology.

It is indeed paradoxical that self-styled democracies support and justify a similar emergence of social *primitivism*. An illusion of this

---

104 Constitutional Council. It is the highest constitutional authority in France, tasked with ensuring that proposed statutes are in accordance with the constitution.

IDEOLOGICALLY DISSIDENT STATEMENTS

kind is the product of the ruling ideologies, which, because of their overconfidence, are no longer capable of analysing reality.

If current demographic and immigration trends continue this way, with a further expansion of Afro-Asian population groups, and Islam continuing to grow increasingly influential –ultimately aiming for its adherents to become the majority, something which few people understand – the future of democracy will be compromised. Gradually, society will be pervaded by authoritarian, fanatical, anti-secular and anti-civil values, with multiracism as a final touch: the latent civil war between communities.

Part of the Left knows all this well, but for them to admit it would be to acknowledge their contradictions and intellectual weaknesses. Most importantly, it would mean going against the multiracial dogma. Out of unconscious racism, the assimilationist Left sees all human beings as neutral and malleable atoms, ignoring their origins. It fails to understand that even after many generations the *ethnic past* endures, like a sort of anthropological atavism. These unrepentant individualists do not grasp that while education may shape an isolated individual, it will never succeed in transforming the values of the ethnic and religious communities that are setting themselves up on the European soil *en masse*. 'Democrats' will face a brutal awakening.

Within the European tradition, democracy – which is to say *the rule of accepted order*, which could also be termed nomocracy or the rule of common law – can only endure provided its citizens share a common and almost innate outlook they have inherited.

It may be that an *authoritarian interlude* will be needed.

4

FOR A TWO-TIER WORLD ECONOMY

Two Ideas in Crisis: Progress and Growth

'Progress' is clearly a dying idea, even if economic growth may be continuing. Yet, no one is really deriving the right conclusions from this. People no longer believe that 'tomorrow will be better than today, just as today is better than yesterday' thanks to technological and scientific advancements and the alleged educational and moral improvement of humanity – the dogma promoted by Auguste Comte and the French positivists[1] – as well as the spread of 'democracy'. Evidence is mounting that 'growth', this measurable mockery, does not actually lead to any objective increase in well-being. The decline of the secular eschatology inherited from Christian messianism is a hard blow for the egalitarian worldview, for it erodes the very philosophy of history on which the latter is based.

Some people believe we are being offered an opportunity here: that we are entering an age of greater clarity and wisdom. Why – they reason – should the end of the myth of progress stand in the way of real improvements and more intelligent forms of progress? Why should it go against the pursuit of equality? These objections, which are frequently raised by members of the 'New Left',[2] are misguided: for progressivism, this pillar of egalitarianism and one of its chief expressions, once served as a global belief and part of its secular religion. A collective

---

1 Auguste Comte (1798-1857) was one of the founders of positivism. Positivism holds that the only knowledge which can be considered reliable is that which is obtained directly through the senses and via the (supposedly) objective techniques of the scientific method.

2 The New Left is the name given to a loose confederation of Leftist movements which emerged in the 1960s and '70s in an attempt to develop an alternative to Communism that was less authoritarian. In some cases, these movements also sought to reform society through the use of existing democratic institutions rather than through outright revolution.

ideal cannot be 'fiddled with' like an economic plan. Deprived of its quasi-religious basis – belief in progress as a historical necessity – the present civilisation has started its decline. But of course it will take an oil tanker whose engines have stopped running some time to come to a complete stop before it starts drifting off towards the rocks...

Historicism vs. Progressivism

The question we must ask then is: with what can 'progressivism' be replaced?

The failure of liberal capitalism to attain its goals of equal justice and prosperity for all, and the collapse of the Communist dream, which pursued the same objectives, have cleared the way for the establishment of a third path. Attempts in this direction have been made around the world by various sorts of authoritarian regimes, all of which have failed – and it is unlikely that fundamentalist theocracies will succeed. Whatever will be the case, this alternative to progressivism can only be based on inegalitarian paradigms, removed from the reductive view of mankind as *homo oeconomicus*.[3] Yet the global intelligentsia, which is still nostalgic for progressivism and whose perspective is twisted by hegemonic thought – the burdensome utopia of egalitarianism – is not ready to seriously consider the prospect of embarking on any new course. Rather, it is clinging to the embalmed body of a dead idea and continuing *as if nothing had happened*.

What has now emerged is not a world unified and nourished by history – the linear and automatic outcome of progress – but rather a chaotic and multipolar one that is undergoing globalisation (through markets and telecommunications); a world that has exploded but is being held together, a disorderly and labyrinthine world that will be increasingly laden with history and 'stories'. The ascending line of progress, which was meant to lead to the redemptive eschatology of a heavenly end of history, is now being replaced by the winding, unpredictable and mysterious flow of this very same history.

The Collapse of the Paradigm of 'Economic Development'

An intellectual revolution is taking place: people are starting to perceive – without daring to openly state it – that the old paradigm according to

---

3   Latin: 'economic man'.

which 'the life of humanity, on both an individual and collective level, is getting better and better every day thanks to science, the spread of democracy and egalitarian emancipation' is quite simply false.

This age has come to an end. This illusion is dead and gone. This advancement (which some, such as Ivan Illich,[4] had already questioned) lasted just over a century. Today, the perverse effects of mass technology are starting to make themselves felt: new resistant viruses, the contamination of industrially-produced food, shortage of land and a downturn in world agricultural production, rapid and widespread environmental degradation, the development of weapons of mass destruction in addition to the atomic bomb, etc. – not to mention the fact that *technology is entering its Baroque age*. All great and essential inventions had already been made by the late 1950s. Later enhancements constitute not so much concrete improvements as additional refinements of little use, like decorative touches added to a monument. The effect of the Internet will be less revolutionary than the telegraph or phone: for it only enhances a pre-existing universal communications system. Technological science conforms to the '80-20' law of energy: initially it takes 20 units of energy to produce 80 of power; but then it takes 80 units of energy to produce only 20 of power.

A possible objection: are we not pessimistically exaggerating the negative consequences of global progress and growth?

The answer to this is no. By contrast to the widely echoed suggestions made by French intellectual Jacques Attali, humanity as a whole has nothing to gain from things like the economic boom in Asia: for the price the older industrial countries would have to pay in terms of an increase in competition would be huge. In any case, this growth will not continue for long: it is becoming difficult to manage – it will have an environmental impact and cause massive socio-political, as well as strategic, problems. Catastrophe itself – not the will of governments – will bring change to the current economic system.

The few positive effects global economic growth brings are actually transient and fragile, and laden with momentous consequences.

In the global spread of technological science, each step forward implies one step back. So life expectancy is on the increase (although it

---

[4] Ivan Illich (1926-2002) was an Austrian philosopher and Catholic priest. In his books he accused many of the major pillars of modern society, such as education, medicine and industry, of what he termed 'counterproductivity', which is when institutions end up impeding the very goals they were meant to attain.

is stagnating if not falling in many countries), but does this mean people are living in greater harmony and with less anxieties? More and more methods of mass destruction – such as nuclear, bacteriological and genetic bombs – are being developed. Agriculture is improving, but ultimately the return of famines is threatening an over-crowded humanity, which inflated thanks to the fall in mortality. We must now face problems such as soil erosion, the destruction of the tropical rain forests, the decrease in arable land, and the depletion of fishing resources.

It will take twenty or thirty years for the pernicious effects of growth to manifest themselves, but after a deceptive phase in which living standards appear to be improving (and which is now coming to an end) they will certainly hit hard. The increase in production and trade leads to new forms of cooperation, but also multiplies the causes of conflict and expressions of nationalistic chauvinism – and everywhere feeds the counter-fire of religious fanaticism. Communication is branching out across the world, while solitude plagues individuals and a sense of despair takes hold in communities.

The urban and technological way of life is shared by 70% of humanity, but what it means – particularly in the South – is life in hellish cities, real cesspools of violence and human chaos. Few know that proportionately more people are living in misery and poverty now than before the Industrial Revolution. Health care has improved, but this has led to a demographic explosion and made the new viral diseases, spread by immigration, more resistant. The global level of energy consumption is rising, while environmental degradation is worsening and the threat of environmental collapse mounting. African and Brazilian farmers now have machines to clear the land, but they are destroying their forests, thus paving the way for desertification and future famine. In other words, after a latency period, progress, growth and the unchecked spread of technological science are producing effects opposite to those desired, engendering a world that is much harsher than the one they wished to transform and improve.

The Announced Death of Global Economic Development

A serious objection must now be addressed: that we cannot possibly prevent poor or 'developing' countries from pursuing industrialisation, striving to attain wealth by all available means and following in

FOR A TWO-TIER WORLD ECONOMY

the footsteps of the West and of the 'global religion of GDP[5] growth'. For what an injustice this would be...

Make no mistake: historical dreams and hopes are not based on *morals*, but on *physical limits*. It is the logic of catastrophe that will limit the ambitions of southern countries to 'develop'. These countries, and particularly those of Asia, have yet to become disenchanted with progress. Behind the West in this respect, they still have a positivist approach and are attached to the egalitarian universalism they have just discovered. They wish to imitate the North and have their share of the pie. But alas, it is all too late. The Asian financial crisis was a sign of what is to come. The planet – and hence humanity – would never be able to cope if all of Asia and Africa were to attain the same level of techno-industrial development as northern countries. To believe this is possible is to exhibit the kind of faith in miracles typical of universalism. The mass industrialisation of 'emerging countries' is most likely physically impossible because of the depletion of resources and the destruction of ecosystems. The prophecies made by the Club of Rome[6] will no doubt prove to be correct some fifty years too late.

Already in the 1960s some Africans, such as Credo Mutwa in South Africa,[7] argued that pre-colonial tribal societies – small, scattered and demographically stable societies – were far more pleasant than contemporary African societies, which are complete failures based on a botched imitation and poor grafting of the European model, one totally alien to them. After all, why should everyone want to reach Mars, travel on 500 km per hour bullet-trains, fly in supersonic jets, live to the age of one hundred through transplants and antibiotics, chat online, watch TV dramas, etc.? This *fever* only belongs to certain peoples and groups, and cannot be extended to humanity as a whole.

Should radical structural changes occur, even in Europe and the United States, most of the population would no longer be able to share

---

5   Gross Domestic Product, or the total economic output of a nation within its borders (as opposed to Gross National Product, which measures the total economic output of all enterprises owned by a country regardless of location).

6   The Club of Rome is a global think tank which was founded in 1968 to address the problems faced by humanity. In 1972 its members published a controversial book, *The Limits to Growth*, which held that the rapid increase in the global population combined with dwindling resources would lead to disaster if changes were not implemented.

7   Vusamazulu Credo Mutwa (1921-) continues to write today, and has always advocated the idea that Africans should return to their native traditions rather than attempt to imitate Western civilisation. He maintains a Web site at credomutwa.com. He has also collaborated with conspiracy theorist David Icke.

the techno-industrial way of life. But here another objection must be addressed, one raised by technocrats: that it is possible to control the perverse effects of technology; that we can fight pollution and find new resources if there's a common agreement and willingness to do so.

This is all very optimistic, but it's only empty talk: it will never happen. The system displays an overall consistent logic and will not transform itself. It is literally incorrigible and must be changed.

On the other hand, a new system will affirm itself – and will do so in the chaos. We must take a concrete approach and stop having daydreams based on the intellectual masturbations of sham experts. None of the resolutions made at the summits of Rio and Tokyo, however insufficient in themselves, have been respected. Nature, which we have sought to dominate and control in all of its molecular and viral forms – and the Earth itself – is now reacting with a violent backlash after a quiet period. Collective certainties are giving way to doubts and distress. A new sort of nihilism is emerging, a highly dangerous, because desperate, one, which has nothing to do with the philosophies of decline and the reactionary prophets of decadence, who merely represented the other side of the dogma of progress: attachment to the past. It is now *philosophies of catastrophe* that will take the stage. We are faced by uncertainty, which is casting its disturbing shadow over the very science of technology that we considered predictable and governable. Heidegger was right in his opposition to Husserl[8] and the rationalists[9] – and the Jewish allegory of the Golem[10] was a most apt one.

Towards a 'Fracture of Civilisation'

What new ideologies or forms of social, political and economic organisation could replace the pursuit of progress and individualism? Are we to return to theocracy, as many Islamic countries would like to suggest? The first thing to note is that non-progressive ideologies that reject

---

8   Edmund Husserl (1859-1938) was the founder of the phenomenological school of philosophy, which was the predecessor of existentialism. Phenomenology has been defined as an attempt to apply the objective methods of science to the study of consciousness, which is viewed as the basis of existence. Heidegger, however, although he was Husserl's student, came to believe that consciousness is only a by-product of existence, which is the actual ground of being.

9   Rationalism is the belief that all of human experience can be understood by reason, and that all phenomena can be explained using the tools of mathematics and science alone.

10 In Jewish lore, a Golem is a human-like creature made from dead matter and given life through mystical powers. In some Golem stories, the Golems get out of control and end up destroying their creators. This is seen as a direct forerunner of the many science fiction stories about intelligent machines turning against their masters.

egalitarianism are not necessarily unjust, cynical or tyrannical. It is the egalitarianists who, conscious of the failure of their plans for justice and humanitarianism, wish to portray their enemies in a diabolical light. New inegalitarian worldviews must prove *concretely anthropophilic* rather than *ideally humanitarian* (like egalitarianism). This end of progressivism clearly also coincides with that of Hegelian rationalist Idealism.[11] Inordinate, irrational, anti-scientific and anti-industrial ideologies are already spontaneously spreading across the world – something which has worried the signatories of the Heidelberg Appeal.[12]

We should, however, resist the temptation to believe that industrial cultures will disappear and be replaced by cultures based on magic.

Technological science will continue to exist and develop, *while acquiring a new meaning and ceasing to be informed by the same ideal*. Global economic growth will soon clash with physical barriers. It is physically impossible to fulfil the ideal of progressivism: the spread of techno-scientific consumer culture to ten billion people. When this dream has faded, another will emerge. According to a scenario I would cautiously envisage (one at any rate far less unrealistic than endless and widespread economic growth in the context of either a world state governed by the United Nations or of a fragmented planet), the following three elements will coexist: *globalisation*, the end of statism and the *collapse of civilisation* worldwide (something which will be passively endured rather than consciously chosen). People preserving a techno-scientific and industrial way of life (yet driven by values other than those we have today) will coexist with people who will have reverted to traditional societies, possibly based on magic, irrational, religious, pastoral and neo-archaic ones with low levels of energy use, pollution and consumption.

Traditional Economies are Not 'Underdeveloped'

Progressive thinkers will retort that what I have just suggested implies organising a sort of voluntary underdevelopment with gifted people consuming available resources above and ungifted people vegetating below.

---

11 Hegel regarded consciousness as something determined by historical reality, which is governed by the laws of dialectics. He asserted that history was a process of humanity developing towards ever-greater states (both political and personal) of freedom, thus paving the way for the utopian ideologies of the Twentieth century.

12 In 1992 a statement signed by many scientists was released to coincide with the environmental Earth Summit. It was a plea for world leaders to avoid advice from irrational or pseudoscientific circles, especially those upholding a call for a return to nature, and other groups hostile to the aims of science and technological progress. It further stated that science, technology and industry are the best means for the 'indispensable tools of a future shaped by Humanity'.

This idea of underdevelopment is both stupid and unjust: it was invented by progressivism in order to argue that the industrial way of life is the only truly human and permissible one. Traditional rural societies not based on technology are not at all barbarous and 'underdeveloped'. According to an inegalitarian and organic worldview, many 'development axes' exist – not just one. True 'underdevelopment', or more correctly true barbarism, is caused by progressivism: consider all the casualties of the industrial way of life, who, for a mirage, have abandoned traditional societies with low demographic rates to join the overcrowded megalopolises of southern countries, real urban hells. Besides, the members of traditional societies where little money circulates are neither 'poorer' nor less happy than New Yorkers or Parisians with all their modern conveniences, even if they may not have health care that is as good and have lower life expectancies. It should also be noted that the socio-economic fracture that is likely to take place in the Twenty-first century will not be the product of any intentional planning, but rather something imposed on humanity by catastrophe and the chaotic collapse of the present system.

But how can different types of society be made to coexist? Won't those below wish to imitate those above and 'develop'? Not necessarily: because on the one hand the failed attempt to globally extend industrial society and technological science will be remembered as a dark age (as Communism is today); and because on the other these neo-traditional communities will be pervaded by strong irrational or religious ideologies sanctioning their modes of life. Those who will preserve the techno-scientific way of life will be perfectly capable of living within a global economic system, albeit one not as vast in terms of production and trade as the one we have today, and hence less polluting – for it will only concern a minority of people. This minority will be driven not by the eschatology of progress, but rather by necessity born of *will*.

A Techno-scientific Economy is the Only Viable One in an Inegalitarian and Non-universalist World

After the inevitable catastrophe that will mark the opening of the Twenty-first century, once the stupid celebrations for the year 2000 are over, it will be necessary to pragmatically plan a new world economy, with a spirit free from all utopias and impossible ideals and from all

FOR A TWO-TIER WORLD ECONOMY

will to oppress or colonise the part of humanity that will have reverted to neo-traditional societies. The prevailing historical outlook will no longer be progressive idealism, but one based on a realistic, *concrete, adaptable and unpredictable* view of reality, nature and man. Voluntarism, the ideology of concreteness and the possible, is opposed to the idealism of contemporary global civilisation, which is based on the abstraction of unachievable goals. Techno-scientific and neo-archaic areas will share an inegalitarian and naturalist worldview: one informed by rationality in the case of the former, and by irrationality in the case of the latter.

Clearly, many will fear that the death of the ideal of progress and the new order of the world will bring an end to rationality and destroy both science and industrial production, thus setting back the whole of humanity.

It is a common misconception, however, that technological science naturally rests on progressive and egalitarian foundations. This is not true: the end of progressivism – with its dream of globally extending industrial consumption – does not imply the dismantling of technological science and the condemning of the scientific spirit. Technological science has been perverted by the egalitarian universalism of the Nineteenth and Twentieth centuries, which has sought to extend its influence beyond all reasonable limits.

Those who will continue living in a global techno-scientific civilisation, albeit one of limited reach, will be driven by values other than the consumer frenzy, universalism and widespread hedonism of the ideology of progress and development.

This will not be difficult, as the foundations of science and technology are actually inegalitarian (life sciences), poetic and adaptable in an unpredictable manner. True scientists know that advancements can only be made by destroying previous certainties. Rationality for them is a means and not an end in itself. These scientists know that discoveries never automatically lead to qualitative improvements, and that technological experimentation implies the unexpected: increased risks, unpredictability and the opacity of the future. By contrast, in traditional societies the future is predictable, because history is experienced cyclically. Hence, in neo-traditional areas linear progressivism will be replaced by a cyclical view of history, while in techno-scientific ones it will be replaced by an *unpredictable and 'landscapist' view of history* (the 'spherical' and Nietzschean view promoted by Locchi, which was previously referenced). In the latter case, history will unfold as a landscape: like an unpredictable succession of flatlands, mountains and forests governed by no apparent rational order.

The above view of history and destiny brings greater freedom, responsibility and clarity to those who embrace it: for they will have to rigorously analyse the true nature of reality and the times, free from utopian reveries and conscious of the unpredictability of things; they will have to apply their will to the implementation of their project – the *ordering* of human society in such a way as for it to conform to *justice* as much as possible – acknowledging man for what he truly is rather than what we would like him to be.

The Neo-global Economy of the Post-catastrophic Age

Another question must now be addressed: based on the premise that the two-tier world economy of the future will be a 'globalised' one, how are we to define the notion of 'globalisation' with respect to universalism? Can these notions truly be opposed to one another? Well, yes.

Universalism is a childish concept based on the illusion of cosmopolitanism. Globalism is instead a practical idea: global information and exchange networks exist, but do not concern humanity as a whole! Universalisation is the ambition to mechanically and quantitatively extend one way of life – industrial consumption and urban living – to all of humanity. Universality is perfectly compatible with statism, and egalitarianism is its driving force. Billions of human atoms are here asked to live according to the same rule: the one imposed by the reign of the market. Globalisation, in contrast, refers to a process of the spread of markets and companies across the world, and of internationalisation of the economic decisions taken by some central actors, without the need for universalism: globalisation is in fact perfectly compatible with the idea that billions of men everywhere may revert to traditional ways of life. On the other hand – and this is a crucial point – globalisation is equally compatible with the construction of semi-autarchic blocs (autarchy for wide areas) on a continental scale based on different economic systems.

After the failure of economic progressivism and market universalism, a global economy may well come to light (and even reinforce itself) that will have no desire to envelop the whole of humanity and will only concern an international minority. This is a perfectly plausible scenario for the aftermath of the catastrophe: for technological science and the industrial market economy cannot be abandoned, as they are too rooted and already in the process of becoming global. But the idea of universally extending industrial society to all individual humans will have to be ditched, for it is unsustainable in terms of energy, health and the environment. The 'neo-global' economy in the aftermath of the catastrophe will certainly be *global* in its networks but not *universal*. The intrinsic inequality of this new economic system will help bring environmental destruction to a halt and restore what has been destroyed – thanks to its low level of energy consumption – and improve the quality of life of all peoples.

Make no mistake: the GDP of the world economy will fall considerably, like a deflating balloon.

One may object that this fall in the global GDP will dry up existing financial resources and make certain investments impossible because of the 'loss of scale' that will have occurred (as the industrial economy will only concern a fraction of humanity, markets and demands will undergo a sizeable contraction). To reason along these lines, however, is to forget that the new economic system will have freed itself from two considerable burdens: firstly, the substantial cutting down of pollution levels will reduce the huge number of external diseconomies,[13] with all their costs, and the burden of having to lend money to 'developing countries' will also have been removed (as the goal of developing these countries will have been abandoned altogether); secondly, the expenses related to state welfare will drop as most of the massive social investments that are currently being made will disappear, as they will have become superfluous given the return to a neo-medieval economic model based on solidarity and proximity.

Clearly, another solution might be envisaged: keeping universalism and persuading rich countries to lower their living standards and energy consumption levels in such a way as to preserve the environment, share wealth with the poor, and balance the industrialisation of 'emerging countries'. According to this shrewd and logical perspective embraced by environmentalists, the solution would lie in more egalitarianism rather than less...

---

13 Diseconomies are forces which increase the costs of production for companies.

The above suggestion, however, proves to be an utterly idealistic and inapplicable one. Rationality is never what matters in history. Can we really imagine Americans spontaneously giving up their cars and accepting to pay double the amount of taxes to help the countries of the South? This said, in a scenario of the economic fracturing of the planet, wide areas and sections of the population *within the industrial countries of the North* could perfectly well revert to traditional forms of economy with low levels of energy consumption and subsistence farming.

An Inegalitarian Economy

What it is important to grasp is the fact that technological science has had devastating effects because it has been driven by the egalitarian ideology of universal progressivism, not because of any intrinsic shortcomings – as Right-wing traditionalists and dogmatic environmentalists believe. The techno-industrial model is now collapsing under the weight of disenchantment because it has been extended beyond all reasonable limits and has been fancifully credited with the miraculous ability to bestow a whole range of blessings. But actually, *by its very nature technological science is something that only tends to concern a minority of the human population*: for it is too energy-consuming for it to be greatly extended.

Clearly, do-gooders will accuse the above theses of promoting widespread *exclusion*. But this is merely another quasi-religious idea that stems from reductionist ways of thinking and the belief that it is morally legitimate to extend present developments to everyone.

Actually, the 'exclusion' of neo-traditional communities from the techno-scientific sphere would coincide with the exclusion of the latter from the neo-traditional world. We should do away with the prejudice according to which techno-scientific societies are more 'developed' than traditional ones. This myth of the savage is an implicitly racist one.

According to the scenario that can be envisaged on the basis of the aforementioned suggestions, neo-traditional communities would in no way be inferior or underdeveloped ones. On the contrary, they would conform to the rhythm of a different kind of civilisation, one no doubt superior to that of today. This inability to free oneself from the dogmas and paradigms of progressivism and egalitarianism, and to envisage different socio-economic solutions, plagues Western intelligentsia as a whole.

Pascal Bruckner,[14] for instance, in an article published in *Le Monde*,[15] starts off by noting the contemporary disenchantment with the failing idea of progress and by acknowledging the pernicious effects of the global spread of technology. But then he adds the following naive comment: 'In contrast to what was hoped for in the Eighteenth century, technological progress is never synonymous with moral progress. Still, a guideline for action exists: the democratic values inherited from the Enlightenment, secular versions of the messianism of the Gospels and the Bible.' What he means by this political cant is: in order to counter the perverse effects of the technological progressivism we have inherited from the Enlightenment, let us return... to the philosophy of the Enlightenment. What ideological idiocy! Bruckner fails to realise that it is precisely the progressive and egalitarian universalism of the Gospels, strengthened by Protestant ethics and the philosophy of the Enlightenment, that has led to the global spread of technological science beyond all reasonable limits through unsustainable growth – an engine out of control – when it was instead necessary to restrict the use of technology to certain areas.

Techno-science as an Esoteric Alchemy

Here's another question: could it be that in envisaging and advocating this socio-economic model an attempt is being made to turn science and technology into confidential matters, something like alchemical formulas reserved for a minority capable of mastering them? Well, this is indeed the case. *Technological science must be decoupled from the rationalistic outlook...* and freed from the egalitarian utopia that seeks to claim it for the whole of humanity.

In a post-catastrophe scenario in which people have experienced the dangers wrought by an unchecked spread of science, technology and the industrial economy, as well as the harmfulness of unrestrained information exchange (excessive communications), it is not unlikely that we shall witness a return to an *initiatic and quasi-esoteric view of technological science*, aimed at protecting humanity from the risks

---

14 Pascal Bruckner (1948-) is a French social commentator who is best-known for his opposition to multiculturalism. Several of his books have been translated, such as *The Tyranny of Guilt: An Essay on Western Masochism* (2010).
15 *Le Monde (The World)* is widely considered the most important daily newspaper in France.

posed by the epidemic, massive and unchecked spread of technology. The ideal would be for this techno-scientific civilisation – a high-risk civilisation, yet one intrinsically linked to the spirit of specific peoples or minority groups scattered around the world – to only be embraced by some people and thus remain esoteric. Technological science cannot be a mass phenomenon – an 'open' phenomenon. The planet rejects this prospect, which is only viable for 10 to 20% of humanity. Let some experience the natural wisdom and certainty of the reproduction of their species, of cyclical time, of rural or agricultural well-being in stable traditional societies; others, the undertakings and temptations of a global and historicised world. For some Guénon,[16] and for others Nietzsche.

---

16 René Guénon (1886-1951) was a French writer who founded what has come to be known as the traditionalist school of religious thought. Traditionalism calls for a rejection of the modern world and its philosophies in favour of a return to the spirituality and ways of living of the past (Guénon himself ended up living as a Sufi Muslim in Cairo). He outlines his attitude toward modernity in *The Crisis of the Modern World*, which is available in English.

5

THE ETHNIC QUESTION AND THE EUROPEAN

AN ARCHEOFUTURIST APPROACH

'They had their faces to the blinding sun. Their lips did not move, but their gazes were threatening. They did not shout like the enemy to give themselves courage. They slowly lowered their lances. The Spartans fearlessly advanced against the countless but terrified Persian ranks.'[1]

To my Greek friends and to Jason Iadjidinas, in memoriam.

Anthropology is the Foundation of History

The ethnic question, along with the environmental, will be one of the most serious challenges humanity will have to face in the stormy century of iron and fire that awaits us. It primarily concerns Europe and, within Europe, France, which is undergoing mass demographic colonisation from other continents – a phenomenon whose magnitude and consequences media and political leaders are seeking to conceal.

The ruling ideology is based on one central dogma: that 'the ethnic question does not matter'. It is always the same story: in the name of a false love of humanity, scorn is poured upon the crucial concept of 'folk'.

Future historians will no doubt study this amazing phenomenon which, as an after-effect of colonisation, has been affecting Western Europe and France since the 1960s. In less than three generations, the

---

[1] This is clearly an excerpt from an account of the famous Battle of Thermopylae in 480 BC, when 300 Spartan warriors, plus a small number of warriors from other Greek city-states, successfully held off an invasion by the much larger Persian army of Xerxes for three days. However, I am unable to identify which source this specific quote comes from.

ethnic substrate of these lands has been radically altered. Surely this should be of interest! Instead, it is only considered a secondary matter by the petty, inglorious princes who pretend to be governing us.

We would do well to read the essay by the Black American sociologist Stanley Thompson published by Boston University Press in 1982, *American Communities*. The author here attempts to evaluate the contribution made by each ethnic community to American society in terms of its 'mentality'. The conclusion of this rather iconoclastic book is that on account of their 'managerial wills', 'honesty in business' and 'pride', Germanic immigrants contributed far more than the English, Scottish, Welsh, Irish or any other immigrant group to strengthen the American imperial republic. The author rather sternly notes that in growing increasingly Hispanic – or more precisely Mexican – the United States will change its ethno-cultural foundations and in the long term possibly enter a phase of decline, in terms of 'objective' power, compared to India and China. The reading offered by this Afro-American and Germanophile intellectual is no doubt an incomplete and exaggerated one, yet it also contains much common sense: for Thompson realised that the basis of a civilisation and the destiny of a given culture are not sheer mechanical facts depending on economic organisation alone, but rather on things that have human and *organic* roots – which is to say, cultural and ethnic roots.

Shlomo Shoam, who was Chair of Philosophy at the University of Ramat Aviv in Israel in the 1980s, shared the following confidential remark with me during one of the Athens symposiums: 'The economic and military power of Israel and its safety in the face of Arab countries rests on its "Sabras" – Ashkenazi immigrants from Europe.'[2] The primary foundation of history is anthropology, which determines cultural behaviour.

The Plan for Bringing 'Ethnic Chaos' to Europe

The ethnic question today is *taboo, and hence crucial*. After a long period of migratory stability, Europe – and France in particular – is now experiencing mass immigration from Africa and Asia, which is changing the ethnic composition of our land against the will of its

---

2   I am uncertain of the meaning of this statement, since 'Sabra' is a term used to designate Jews born within Israel. Perhaps Prof. Shoam was referring to the Israeli-born descendants of the Ashkenazi immigrants.

THE ETHNIC QUESTION AND THE EUROPEAN

native population and in contempt of the democratic traditions we have inherited from the Greek cities, the Roman Republic and Germanic law.

Immigrationists reason that France has always been a land of melting-pots and large-scale invasions. The proof of this? The endless waves of Celts, Germanics, Romans and Slavs that have entered the country. Sure, but these were neighbouring peoples, 'close cousins' to be more precise. France is indeed a mix of almost all the ethnic components of our continent, including the Germanic, but these were all populations with mental structures and forms of behaviour close to our own. For the notion of ethnic proximity, while necessarily bio-anthropological in nature, primarily concerns peoples' proximity in terms of worldview and instinctual attitudes. King Clovis – Kounig Chlodovech, to call him by his name – was assigned the role of Roman consul by Constantinople. Mental continuity thus existed in the land of the Gauls between the Roman and the Germanic worldviews, which were added to the existing substratum of the related Celtic peoples.

It is well-known that, from an ethnic point of view, France is a synthesis of European peoples. Immigrationists justify the massive flux of immigrants from Africa and Asia by arguing that France has always been a land of 'miscegenation' and hence nothing has changed – that we're merely continuing our tradition and that there's nothing to worry about. Actually, the 'miscegenation' in question only occurred between European peoples. The Germanic 'invaders' – the most commonly invoked culprits – were not quite as invading as one is led to believe; for after all, they were probably already present in the land of the Gauls prior to their alleged 'invasion', sharing a culture that was very similar to that of the Gallo-Romans. The real invasions are not those that occurred in late Antiquity, but those we are experiencing today.

Here's another sophism used by the immigrationists: the idea that the percentage of foreigners in the French population today appears to be much the same as that of the year... 1930. To believe this is to ignore the mass naturalisation of immigrants that has occurred and – most importantly – the fact that thanks to the aberrant law of ground (*jus soli*), millions of 'young people' of Afro-Asiatic origin, who do not see themselves as being French at all, are indeed regarded as such by law. These people reason in ethnic terms – unlike Parisian intellectuals.

The mixing that took place in the land of the Gauls, whatever its scale, only occurred among peoples who were cousin-folks from the point of view of anthropology and culture, as well as linguistics. By contrast, the Afro-Asiatic populations which have moved to our continent since the 1960s, altering its ethnic and cultural composition (the Muslims in France will soon reach 5 million and, from around 2005, Islam will be the most practiced religion in the country),[3] share no anthropological, cultural or even mental proximity to European natives – unlike the Germanic populations with respect to the Romans, Celts or Slavs. What we are witnessing, then, is a break from tradition, not any form of traditional continuity. On the other hand, the 'Germanic invasions' of late Antiquity, like all the other military incursions or flows of immigration that France experienced in one thousand years of its history – at the hands of the Moors, English, Dutch, Spanish, Germans, Russians and Italians –never caused any radical ethnic changes or cultural dichotomies. Hence, when the partisans of immigration compare these intra-European movements to the mass demographic colonisation to which we are being subjected today, they are quite wrong: theirs is merely an intellectual absurdity used to conceal the true nature of what is happening.

With their twisted – and ultimately anti-democratic – reasoning, these people aim to favour the spread of *ethnic chaos* in Europe, while concealing its reality. Let us not forget that the immigrationist lobbies are headed by Trotskyists, whose irrational and hidden feeling has always been *hate* for European ethno-cultural identity.

Besides, these internationalists are supported in their plans by ultra-Liberalism of American inspiration. The geopolitical goal of the United States – and we can't really blame them for playing their cards – is to dominate the continent of Europe, destroy its ethno-cultural identity and take over its markets and techno-economic resources.

No doubt, France had already experienced a series of immigration fluxes in the early Twentieth century – at the hands of Spaniards, Italians, Portuguese, Poles, etc. But again, these were all peoples from areas not far away: Catholic folk who spoke related languages and even had a sort of shared historical memory. Henry III was 'King of Poland,'[4]

---

[3] As of this writing, Islam is still believed to be second to Catholicism in terms of practice in France, reckoned at approximately 10% of the population, although it is difficult to gauge how accurate these estimates are. The statistics released by the Church itself indicate that practice among French Catholics has been dropping rapidly.

[4] King Henry III (1551-1589) was King of France from 1574 until 1589. Prior to that, he had been elected to be by the noblemen of the Polish-Lithuanian Commonwealth, which had been formed in 1569, and held the titles of King of Poland and Grand Duke of Lithuania beginning in 1573.

and all of European history is but an assemblage of transcontinental 'fragments of memory'. French history cannot be understood without constant references to Germany, Italy, Russia, England, Spain, etc.

These intra-European migrations (which in any case took place on a far more limited scale than contemporary migrations from Africa and Asia) may be compared to migrations within North Africa or from continental China to the country's maritime areas. A degree of 'mental distance' certainly exists between contemporary Flemings or Germans on the one hand and Greeks or Sardinians on the other, but it is considerably less than that which separates us from the ethnic blocs of other continents.

Can people simply be mixed together, as a cook would mix his vegetables to make a salad?

We should not hesitate to speak up against the crypto-racist ideology of the partisans of unchecked mass immigration.

Immigrationist lobbies – of Trotskyist observance – are perfectly aware of the fact that *multiracial society means multiracist society*: something that has already been noted many times in the present work but which it is worth stressing again and again.

France, Europe and the German Question

I would now like to address two other thorny questions: anti-German sentiment, which reflects a repressed feeling; and then the following: why still worry about ethnic problems and immigration today, in the age of the Internet and globalisation? Is this not an outdated concern? After all, are we not all citizens of the world?

Let us engage in a little political psychoanalysis, without forgetting our sense of humour. Anti-German sentiment among the French is the product of three European civil wars: those of 1870, 1914, and 1939. These may be seen as a German 'reaction' to French aggression under Louis XIV[5] and Napoleon. Luckily, this feeling has receded thanks to the building of Europe and Franco-German cooperation, which was

---

However, he was also in the line of succession for the French throne, and when his predecessor Charles IX died in 1574 he abandoned the Commonwealth to become King of France.

5   King Louis XIV of France (1638-1715) was King in 1700, when Charles II, the last Habsburg King of the Spanish Empire, died and designated Louis XIV's grandson as his successor, who was also in the line of succession for the French throne, meaning that a unification of France and Spain became a possibility. The Holy Roman Emperor formed a coalition of nations to restore the Habsburg entitlement to the Spanish throne, leading to the multi-continental War of the Spanish Succession. Much of the fighting took place in Germany.

initiated by de Gaulle. Still, (in both France and Great Britain, countries with strong Germanic roots) anti-German sentiment continues to exist in an embryonic form, as a potpourri of dumb clichés, unconfessed hatred, repressed resentments and fantastical fears: 'German, what a ghastly language!' (what about Hölderlin,[6] Rilke[7] or Nina Hagen?);[8] 'Those Germans want to take over Europe!'; 'Deep down, they're still Nazis...', etc. The silly jokes cracked about the Belgians (whom the French in their collective unconscious perceive as 'francophone Germans') or Swiss Germans are indicative of the same fantasy – one that was first engendered during the European civil wars, when people enjoyed drawing contrasts between a distinguished and refined Celtic-Roman French 'race' on the one hand and a simple-minded, brutal and barbarian German one on the other.

German journalists and intellectuals are also responsible for this depreciation of their own ethnicity and culture, for they never cease explaining Hitler's dictatorship as a product of typically Germanic psychological traits. This is a form of masochism and self-flagellation. Are the Russians collectively blamed as a folk for the crimes of Communism? This permanent suspicion of all that is German, of which the guilt-ridden Germans themselves are victims and accomplices, weakens the cultural power of our continent, for it neutralises the Germanic component of the European genius.

Insidious anti-Germanic sentiment, which still pervades French society, is more of a socio-cultural thing and is not directed towards Germany as such. This is quite normal: one does not mock one's 'number one client'. In the issue of the newspaper *Libération* published on 9 December 1997, a 'sociologist with fieldwork experience' learnedly argues that the fact that 'young people' in the Alsatian city of Mulhouse are wrecking local buses can be explained on the basis of the 'racist' attitude of local bus drivers. And of what does this 'racist attitude' consist? In dirty insults aimed at 'young people' born of immigrant parents? Nope! 'These people speak Alsatian with one another and this is perceived as an act of provocation', our comic-opera sociologist explains. In other words, using one's native Germanic language in one's

---

6 Friedrich Hölderlin (1770-1843) is considered one of the greatest poets of the Romantic era. His work has been translated.

7 Rainer Maria Rilke (1875-1926) was an Austrian regarded as one of the greatest poets of the Twentieth century. He wrote in German. His work has been translated.

8 Nina Hagen (1955-) is a German rock star who has had international fame since the 1970s.

THE ETHNIC QUESTION AND THE EUROPEAN

own country is intrinsically perceived as a racist provocation. What a nightmare! Actually, it is the explanation provided by this pseudo-sociologist which is deeply and naively racist. His slip of the tongue reveals a form of racism as unacceptable as all other forms of hatred directed against any given folk. For don't racism and hatred commence when one rejects the very notion of folk? This is an extremely interesting example: for ultimately, according to the ruling ideology, everything European and rooted is perceived as being guilty and criminal. Guilty, that is, of being itself (ethnomasochism).

By tradition, culture, heritage, education and outlook I am Latin and Hellenic. I thus feel perfectly comfortable with expressing what Europeans consciously or unconsciously expect from the Germanic spirit, which extends far beyond the borders of Germany. What are the 'ancient' Germanic qualities that have long contributed to shape Europe?

Firstly, a democratic fibre – understood in the etymological sense of the term, as the situating of the will of the people above any judge's decrees, whereby it is this will that is the basis of the law and not vice-versa. Communitarian solidarity is here regarded as more important than socio-economic hierarchies. Respect for women, the keeping of one's word ('frankness'), honesty in business, punctuality, active dynamism, creative inventiveness, skill in collective organisation and scientific rigour: these are all Germanic qualities.

Yet the Germanic soul also has its drawbacks, which is why it should be tempered with the different mental dispositions of its European cousins. Take its Romantic tendency to 'go to the very end of things', which Madame de Stäel[9] so aptly identified in the early Nineteenth century. This excess can lead to both exacerbated nationalism and organised, suicidal and masochistic laxity (e.g., the *Grünen*),[10] to statism as much as anarchy, suicidal militarism as much as suicidal pacifism, self-exaltation as much as self-flagellation, and complete materialism on the part of individual consumerists – *homo BMW* – as much as disembodied and inert spirituality.

The fact remains that the block of Germanic populations lies at the axial centre of our continent (which is currently undergoing a difficult

---

9   Anne Louise Germaine de Staël-Holstein (1766-1817) was a Swiss writer who was also known as a supporter of the French Revolution and a critic of Napoleon. She also wrote on the cultural trends of the time.
10 German: 'Greens'.

process of unification) and contributes to shape many vast regions. The Germanic soul permeates the most dynamic aspects of all European countries. 'Germanic', however, means more than merely 'German'. De Gaulle's plan for European independence, the Ariane rockets, the Concorde and the Airbus are all components of a political project whose cultural essence is Roman (the will to imperial power), while also being informed by Celtic ardour and Germanic rigour and engineering skill.

It was France, a country as Germanic as it is Celtic and Roman, that has benefitted the most from this intra-European ethnic synergy. This geographically exceptional country and crossroads of European peoples is a synthesis of Europe. The problem is that we must now choose a new horizon: France as a micro-Europe or Europe as a macro-France? Not a 'French' Europe, of course, with all the calamities this implies – Jacobin *jus soli*, taxation, bureaucracy and centralism – but one different from that based on the chaotic constitution it has given itself today, and which may adopt a political plan, as the French state did for a thousand years. It is interesting to note that it was the French and Germans – the 'Franks of the West' and the 'Franks of the East', to quote the German poet Stefan George[11] – that together with the other Franks, the Belgians, have been the promoters of this great plan.

The European project must be pursued in ways more effective than that crippled and paralytic old dinosaur of the European Union sprung from the Amsterdam Treaty.

The Frauds of Globalisation and Cosmopolitanism – How Tomorrow will be an Ethnic World

Is worrying about ethnic questions not pointless in the age of globalisation? Not at all – it is futuristic: for we are not moving towards the disappearance of the notion of folk, but towards its strengthening.

Both the partisans and the enemies of 'globalisation' are tilting at windmills. Through international trade and exchanges, globalisation had already occurred between the Sixteenth and Twentieth centuries – this is now an established fact. It was first set in motion by Europe with its 'great discoveries', the conquest of America, and colonisation.

---

11 Stefan George (1868-1933) was one of the best-known poets of his age. Although apolitical himself, his poetry and ideas were very influential upon the German intellectual Right. His most famous student was Claus von Stauffenberg, the would-be assassin of Adolf Hitler in 1944. His work is available in translation.

THE ETHNIC QUESTION AND THE EUROPEAN 183

Still, the globalisation of commerce has never been synonymous with ethnic intermingling or with unchecked free trade. We are experiencing globalisation today: this simply means instant communication and the establishment of trans-national communications, as well as strategic, economic, scientific and financial networks. Still, first, globalisation does not prevent the United States from basing only 12.4% of its economy on extra-continental trade; second, globalisation does not prevent France, Italy or Germany from keeping the vast majority of its exports within Europe; and third, globalisation only affects a small percentage of human activities.

What we should be critical of, from our point of view, is rather the champions of globalisation – or, more precisely, *cosmopolitanism*. This term serves not as a means to describe an existing reality, but as a weapon of ideological warfare against Europe, destined to anthropologically flood our continent after having paralysed it politically.

These champions of cosmopolitanism say, 'The people of the Earth are one, so let us intermix.' They would like us to believe that the future of the planet consists in widespread intermixing, and that political and economic frontiers are being eroded. But theirs are only sophisms: this is not at all what's happening. Ethnic homogeneity through miscegenation is not at all waiting round the bend; on the contrary, ethnic blocs are growing stronger. Only Europe and North America are being subjected to immigration. Only Europe and North America – or, rather, their intelligentsias – believe and make others believe in the inevitability of a global melting-pot. Just as Marxism made people believe in the scientific inevitability of the rise of internationalist socialism, globalisation represents a central component of the cosmopolitan ideology, which is so wisely explaining how we are 'historically' forced to accept the mass influx of Afro-Asiatic immigrants and to relinquish our ancient anthropological and ethnic identity as Europeans.

Now, globalisation and immigration do not concern the rest of the world. It is an intellectual deception to argue that globalisation is a world-wide phenomenon reflecting the course of history. What is real, by contrast, is the mass demographic colonisation we are being subjected to. China, India, Africa and Arab-Muslim countries are no longer intermixing: they are exporting their blood, while preserving themselves as closed blocs. They are conquering us (partly as a form of revenge, as previously argued) through a method of *infiltration*, which is far more effective than open military invasion – for it won't trigger any immediate reaction and revolt.

Still, a concrete medium-term risk exists of ethnic civil war in Europe, should the latter rediscover its identity and lost homogeneity. This would take the form of a *civil revolt* on the part of native Europeans, which might be triggered by the aforementioned convergence of catastrophes. The dumb pacifism of the immigrationists and their dreams of harmonious intermingling will lead straight to war. But so much the better: stupid ideas are always overthrown by hard facts.

Should We Abandon the Idea of a 'French State' in Favour of a European Federation?

I have no faith in the idea of 'world citizenship'. On the other hand, I've never been much attached to the French state, which is essentially a high-tax, centralised and unrepentantly Colberto-Socialist[12] entity, a leech sucking the blood of the Gauls and a cause of world wars in the past. Attached to the untenable *jus soli*, in the long run it will destroy what it has been entrusted with defending: the French people. The *jus soli* was easy to assert, like one of those gratuitous and romantic slogans from the age of the Revolution ('All men have two fatherlands: their own and France'). Ideologues treat the term 'French' as a political concept, while the people have always understood it as an ethnic notion. At the time when it was formulated, there were no mass immigration flows, and so there were few risks involved in promoting utopias.

It is a shame that many of those who claim to be 'attached to France', such as the Front National for instance, are not choosing the path of a European federal empire, but rather insist – out of nostalgia and romanticism, no doubt – to show micro-nationalist attachment towards the French state. These people fail to realise that the French state is intrinsically destructive for the ethnic identity of the French people, and that *it cannot be changed in its essence*, for it has proven incapable of protecting us from unrestrained immigration. Would a federal European state be more capable of doing so? I believe it would, provided it is exactly the opposite of the European state that is currently being built.

---

12 This refers to the economic policies of Jean-Baptiste Colbert (1619-1683), who was the French Minister of Finance under King Louis XIV from 1665 to 1683. His doctrine, which has come to be known as Colbertism, was built on the premises that the wealth of a nation should primarily serve the interests of the state, and that state intervention should be used to ensure this.

THE ETHNIC QUESTION AND THE EUROPEAN

People and groups like the Front National are actually quite right to speak out against the Europe Union based on the Amsterdam Treaty, a bureaucratic and apolitical monster which contributes to unemployment with its free-market ultra-liberalism, encourages immigration through its pseudo-humanistic ideology and the utter porosity of its outer frontiers, is responsible for the desertification and environmental pillaging of the countryside, limits civil democracy with its pre-totalitarian technocratic tendencies (EU 'directives' would be worthy of the Gosplan),[13] and in all strategic and commercial matters yields to the diktats of its American overlords – for the EU is only an administrative organ with no sovereignty at all.

There is little doubt that, through the fool's bargain of the EU, nation-states are foregoing their sovereignty and replacing it with a complete void – with 'nothing' at all: a legal dinosaur devoid of any political will and utterly incapable of defending us. Yet, the alternative to this is not a return to the states under siege that existed before the War, nor a Europe based on the 'mutual understanding of nations', as envisaged by Talleyrand.[14] The solution to help us defend ourselves must be a radical one: a 'good' federation (one I believe should be based on autonomous regions) capable of imposing itself as a genuine state and exercising a weighty influence on the international scene as a real world power. A federation of this kind could only emerge after a shock, once the pseudo-federation we have now has shown all its impotence and noxiousness.

I believe the right strategy would be to lead a *revolution within the European Union*, in such a way as to radically transform it – and not make a backward-looking return to the nation-states system, which in any case would be incapable of defending us. In history, only *structural changes* can reverse what exists and bring revolutions about – not *circumstantial changes*.

France – like Germany – is finished as a political entity. Europe must take its place. Like the late Middle Ages, ours is a difficult age of

---

13 Gosplan, or the State Planning Committee, was the body in charge of economic planning in the Soviet Union.
14 Prince Charles Maurice de Talleyrand-Périgord (1754-1838) was a French diplomat who began his career under Louis XVI, continued through the French Revolution and the reign of Napoleon, eventually turned against Napoleon and aided his opponents, and then under the first three French kings after the restoration of the monarchy. In this context, however, Faye is referring to Talleyrand's participation as the French representative at the Congress of Vienna in 1814 which sought to restore order to Europe after the Napoleonic Wars. The Congress set up the system of international relations in Europe which lasted until the First World War.

interregnum, albeit in an inverse sense. France will survive, but not as a legal entity: rather, as a culture in the Germanic sense of the term.

The only hope for salvation in this dark age of ours lies in the attempt to build a federation – the great federation Nineteenth century visionaries had foreseen: the United States of Europe. A federation of this kind would be capable of standing up to the American one, of creating a protected and self-centred continental economic space, and of curbing the rise of Islam and demographic colonisation from Africa and Asia. As history is gaining momentum, if Russia were to join us we could start working on the tremendous project of building Eurosiberia.

Despite all its defects, I believe the present European Union will be the prelude to a genuine federation, according to a dialectic process: for when catastrophe hits, the present Union, in its impotence, will have to undergo revolutionary change (this, and not any dangerous restoration of the nation-state model is the path we will have to pursue).

The slogan 'An independent France within a strong Europe' is a utopia and a contradiction in terms, for:

1) a strong Europe cannot be based on an agreement between twenty independent nations;
2) independent nations that will not agree to transfer their sovereignty cannot serve as the basis for a strong Europe;
3) a powerful Europe, in my view, cannot but derive from the federation of autonomous European regions, as the great differences in size between European nations prevents the building of any viable federal and political union (as shown by the current, stupid attempt to do so).

For this reason, we must approach the European Union of today with Machiavellian cynicism in order to *subvert* it from within. Alain de Benoist has made exactly the same analysis as me in promoting the idea of a European empire, rejecting the French Jacobin model, and denouncing the shortcomings of the bastard Union we have today. De Benoist has also explained why he voted in favour of the Maastricht Treaty (see *La ligne de mire, II*).[15] Europeans are perhaps in the process

---

15 *La ligne de mire. II, 1988-1995: discours aux citoyens européens* (Arpajon: le Labyrinthe, 1996), or *The Line of Sight, vol. 2, 1988-1995: Addresses to the Citizens of Europe.*

of clumsily laying the foundations for a new state or – to be more exact – a new empire. Like all great revolutions, this is taking place amid scribbles, not a flourish of trumpets. It is letting itself be led, to use Lenin's expression, by useful idiots who are haunted – and this is a sign of the folk unconscious – by a badly formulated intuition (according to the logic of political suppression Pareto described): the development of a macro-continental defence strategy in the face of the increasing threat posed by outside peoples – the 'giant hedgehog' strategy.

Make no mistake: the present European Union is far from perfect, as are all great historical works in the making. Nothing takes place according to the visionary scenarios drawn by intellectuals, for 'all art is suffering' – as Nietzsche put it. But it is precisely because this Union is imperfect that we should jump on the historical bandwagon to correct it and pave the way for revolution.

Again, the dialectical passage from the impotent and oppressive European Union we have today to the federation I have envisaged could only take place through the mental shock engendered by catastrophe (it is worth bearing in mind that the radical change of spirits caused by the defeat of 1940 led to forms of political organisation that had previously been inconceivable). Quite simply, this appalling Union has the simple yet great merit of making *the whole world reason in terms of Europe*. It also has the advantage of assigning a greater significance to regions, the future bricks of a federal empire, which are connected to the kind of ethnic identity the cold and crisis-ridden states of today have lost.

An ideology is powerless unless it enters the arena of debate. If it limits itself to the idea of 'France' it will never have any political influence. The followers of Maurras *made an ideological exit from history* the moment they chose to attach themselves to the old notion of royalism. We must make sure not to make the same mistake by sticking to French nationalism, which is now obsolete. A new container is in the making: the European Union. Let us fill it with what we have. *European nationalism* is the way forward.

Not the Destruction of France, but its Redefinition as 'Gaul'

Is it not quite clear that the republican ideology of the French nation-state is incapable of defending the people of the Hexagon?[16] That French culture and language don't need this state? And that we already have a political entity that has made the formidable decision to adopt a single currency and flag, and which is effectively a state in the making?

Alone in its isolation and accounting for only 0.9% of the world population, France cannot be protected or made dynamic. Already 40,000 Frenchmen have moved to Silicon Valley, near San Francisco, as expats, and have been replaced by as many illegal immigrants with no skills. As for the 'Europe of nations' model, which does not imply any transfer of sovereignty, it would only create an empty shell where the Americans – 'the first European power', as they themselves like to say – would play divide and conquer. In order to affirm ourselves and resist in the difficult century in the making with its vast world blocs, we need an empire, not a diplomatic association of small or medium pseudo-independent nations (which will never reach any mutual agreement) on the obsolete model of the Congress of Vienna of 1815.

Those who believe that an imperial and federal European state would 'kill France' are confusing the political sphere with the ethno-cultural one. Their notion of belonging is a mechanistic and static one. The disappearance of the Parisian state – to call it by its name – would in no way threaten the vigour and identity of the people of old Gaul. On the contrary, it would reinforce them.

To build a future federal (and imperial) European state, the statist French notion of the '*jus soli*', inherited from the Revolution, must disappear. The simple reason for this is that the traditions of the British, Spanish, Germans, Slavs, etc. are closer to the right of blood, and so the French state will have to forego part of its universalising claims. Obstinate attachment to the French Jacobin state – be it from the Left or Right – means paving the way for automatic mass naturalisations. Those who are naturalised, rather than integrating, will never feel French, but will always continue to feel Arab or African. For they reason in ethnic terms.

Unfortunately, there is already talk in Germany today, under the influence of the French Left and out of chronic self-hatred and guilt, of adopting the law of ground. Yet from the point of view of a European

---

16 France is often described as a Hexagon due to its geographical shape.

federation, based on autonomous regions with traditional roots (and no longer mentally dependent upon the disembodied Jacobin ideology and the idea of cosmopolitanism sprung from the French Revolution), places like Bavaria, the Palatinate, Burgundy or Occitania will have returned to *ethnic entities* and will thus find it easier to do away with the present taboos surrounding the right of blood, a right which they would include in their legislations.

The passage to a federal state would not destroy the *physical substance of France*, but rather reinforce it. How? By breathing new life into autonomous regions: Brittany, Normandy, Provence, etc., which would rediscover their individual personalities within a common European home. Within a federal Europe, France would return to be what it is, deep down: Gaul.

For a Democratic and Federal European Nationalism

We must abandon French nationalism along with the shady pseudo-Europeanism of the Brussels Commission[17] and play the card of a third way, European nationalism, in the framework of EU institutions. We must do so with intelligence and by avoiding manifest extremism. How can it be normal for those who have always dreamt of a great Europe to balk at boarding the plane when it is about to take off? Even if they don't like the pilots, shouldn't they have the courage to play the pirates of the air?

I would now like to examine a number of crucial points concerning the way we should shape this nationalist view of a future United States of Europe. Clearly, these are only outlines and suggestions. History shows that all revolutionary thought must be based on a *set programme* – as Caesar, Napoleon and Lenin knew well – until a *collective shock* occurs that, through the wavering and sinking of people's spirits, will enable its implementation. The making and affirmation of new historical entities depends on the meeting of these two notions, which serve as the sperm and egg of history:

We must embrace a genuinely democratic – and no longer bureaucratic – European government with a real parliament and a strong and decisive central power.

---

17 Brussels is the location of the European Commission, which is the governing body of the EU.

We must do away with the national dimension, which is no longer viable (it is ridiculous, for instance, for the presidency of the EU to be assigned to Luxemburg after Germany), particularly now that plans have been made to extend the EU to central Europe. We must then establish autonomous regions or *Länder*,[18] according to an extended German model (Brittany, Bavaria, Scotland, etc.), where general agreement will determine the political will of each federal power and the president of the Union will be directly elected. Regional autonomy would reinforce the ethnic character of the Union, which is currently overshadowed in France by the ideology of the state. Ethno-regional identity is already gaining increasing importance across Europe (in the United Kingdom, Italy, France, Belgium, etc.). This is a 'weighty historical tendency', to use Fernand Braudel's[19] expression. This form of regionalisation must be promoted, not in a vaguely romantic way, but by illustrating its technical institutional advantages. A Union composed of fifteen different states of variable sizes would not be easy to govern. It would be better off to have seventy *Länder*, each protecting its own autonomy and democratically representing the local population, and a de-bureaucratised central government – with Brussels as the capital and 'federal district' of the Union – that would be something more than the present rump-parliament in Strasbourg.

The United States of Europe, an organic assembly of large and highly autonomous regions (some of which would consist of present states, such as the Czech Republic and Ireland), would determine a new world geopolitics and accelerate the course of history. Only in this framework would it be possible for Europe to compete with the dollar, emancipate itself from NATO and negotiate with the United States on an equal footing. Considering human cowardliness, I believe that this order, structural revolution (secretly planned since the end of the European civil wars in 1945) and difficult birth of a new and internationally influential historic entity will profoundly change the outlook of the contemporary French people, who are currently victims of the whims of the Parisian state. We must trust history, which is synonymous with movement, change, and assault.

---

18 'States', which in present-day Germany includes Bavaria and Saxony. Faye is likely referring to the *Länder* of the Holy Roman Empire, however, in which then individual states retained some degree of autonomy under the leadership of the Emperor.

19 Fernand Braudel (1902-1985) was the most prominent French historian of the Twentieth century.

THE ETHNIC QUESTION AND THE EUROPEAN 191

At the same time, it is necessary to envisage a radical redevelopment of the 'Schengen Area'[20] of free inner circulation and consider adopting a 'fortress logic' for the Union.

The future regions must be granted large powers with respect to internal matters (cultural, linguistic, educational, etc.), as a return to regional identity on a European level would only contribute to our common strength. Different but united: for united we stand, divided we fall.

From an economic point of view, we must consider the prospect of establishing a semi-autarchic common European space. Global free trade is not viable. The united Europe of the future must terminate the GATT agreement[21] and adopt a moderate but effective form of continental protectionism. We are numerous enough not to have any vital need for foreign trade, which often also implies dangerous transfers of technology.

In the long term, we must think in Euro-strategic terms. Gorbachev had understood this well: 'Ours is a common home,' he noted.[22] From Brittany to Kamchatka, 25,000 kilometres lie between the shores of Groix and those of Kerinask; but the men are the same, the virtual citizens of a common empire, and ultimately members of the same folk: the European. We can accommodate guests, but not invaders. Gorbachev simply wished to express this intuition: that we are part of the same group of peoples; that we should stop waging war against each other (as in the Yugoslav Wars, the last foolish European war) and unite. Our linguistic differences are only details compared to our ethnographic commonalities. This is the Germanic approach to history as ethnic logic asserting itself against the utopia created by the French Revolution, which has nothing particularly 'democratic' about it (in the Greek sense of the word) but, on the contrary, is strikingly totalitarian.

We would do well to join Russia one day and envisage the future in terms of Eurosiberia. The unpleasant conditions Russia finds itself in today are only a transient and short-term problem. All we must do is

---

20 This refers to an agreement that was signed between 25 European nations in 1985 in Schengen, Luxemburg, which allows for the free passage of citizens from one country to another. It was absorbed into the EU in 1999.

21 General Agreement on Tariffs and Trade. Negotiated under the auspices of the United Nations in 1949, it remained in effect until 1993, although its terms are still enforced by the World Trade Organisation.

22 Gorbachev had used the phrase earlier, but is most famous for using it in an address in Prague in April 1987, in which he was calling for an end to the partitioning of Europe between East and West.

counter the (natural and understandable) will of the United States to control Eurosiberia and lend Russia protection and financial assistance in view of its future strategic and economic reduction to subservience.

Eurosiberia

Celts, Germans, Greeks, Slavs, Scandinavians, Romans, Iberians... or, rather, *we*, the descendants of these peoples, must now see ourselves as part of the same folk and the inheritors of a common land – a vast motherland with colossal resources, both material and human, shaped by a common history. According to the less ambitious hypothesis, this land would stretch from the Atlantic to the Russian borders; according to the most ambitious one (which must always be promoted), it would be identified with Eurosiberia, which may also be taken as a paradigm for the idea of 'Greater Europe': a land stretching from Brest to the Bering Strait, twenty-four times the size of France. This would be the largest unified political entity in the history of mankind, one extending across fourteen time zones. 'Politics is only for those capable of having a broad, very broad view of things', as Nietzsche said.

One of our frontiers would be the Amur River – our border with China. Others would be the Atlantic and Pacific, our borders with the imperial American republic, the leading world superpower but one whose geostrategic and cultural decline has already been 'virally' programmed for the first quarter of the Twenty-first century – as foretold by Zbigniew Brezeziński (an apologist for American power nonetheless). Two other frontiers of ours would be the Mediterranean and the Caucasus, our borders with the Muslim block (which is less divided than is commonly thought). This block will give us no quarter and will probably represent our greatest threat; but at the same time, if we are strong enough, it may represent an excellent partner...

We, the descendents of related peoples, are being offered the chance to share a space that, already in our children's lifetime, may come to embody what Charles V[23] dreamt about but was unable to preserve: 'An empire on which the sun never sets.' When it is noon in Brest it is 2 AM on the Bering Strait (and vice-versa). This is an ideal we can pursue, one of the few remaining ones in this age of pessimism: to build an empire of our own. What a haunting dream! Great plans are

---

23 Charles V (1500-1558) was Holy Roman Emperor and ruled over a vast area of Europe. He was forced to fight several wars against France.

THE ETHNIC QUESTION AND THE EUROPEAN

drawn not with pomp and solemnity, but in the silence of cabinets; and they are implemented by predators on their guard for a historical disaster to happen and make their prey emerge from the undergrowth in panic. The folk unconscious will always be the hardcore stuff upon which the plans of revolutionary leaders will rest.

In human history, the establishment of a Eurosiberian complex would represent a revolution greater than that of the short-lived Soviet Union or even the United States of America. This event of global importance could only be compared with the foundation of the Chinese or Roman empires.

Whatever the reasons explicitly given to justify the process – and which are of little importance – the European family is coming together in its common home. As in the past – like the Greeks against the Persians almost 2,400 years ago[24] – we are uniting our cities to face a vague but already perceivable threat. Greater Europe must be peaceful and democratic, yet autonomous, inflexible and invincible – clearly, in the technological and economic sphere too; for what need does an empire have of being imperialistic? An imperial logic will extend to all peoples of the earth. Each folk in its own land to defend itself from the aggressions of others, effectively managing the destiny of spaceship Earth.

The chaotic event we are witnessing – this disorderly grouping of Europeans, which only awaits to organised – may represent the reconstitution and historical reoccurrence, in a different and larger form, not only of the Roman Empire, with its centre in the Mediterranean, but also of the Holy Roman Empire, with its centre on the vast Eurosiberian plain, which opens onto four seas. Leviathan[25] and Behemoth[26] rolled into one.

A view of tomorrow: from the harbour of Brest to Port Arthur, from our frozen islands in the Arctic to the victorious sun of Crete, from the fields to the steppe and from the fjords to the maquis, a hundred nations free and united, regrouped to form an empire, will

---

24 The Greco-Persian Wars lasted for half a century, between 499 and 449 BC. Although the various Greek city-states were usually at odds with one another, they united against the threat from the Persians. Thucydides and Herodotus wrote the most famous accounts of the conflict.

25 Leviathan is a huge sea monster mentioned in the *Old Testament*.

26 Behemoth is an enormous creature described in the *Book of Job*.

perhaps be winning for themselves what Tacitus[27] called the Kingdom of the Earth, *Orbis Terrae Regnum.*

---

27 Gaius Cornelius Tacitus (56-117) was a Roman senator and historian.

6

A Day in the Life of Dimitri Leonidovich Oblomov[1]

A Chronicle of Archeofuturist Times

Brest, 22 June 2073, 7:46 AM

The Brest-Moscow-Komosomolsk bullet-train left at 8:17 AM. The Plenipotentiary Councillor of the Eurosiberian Federation, Dimitri Leonidovich Oblomov, was running late. He hadn't slept much and had woken up at the last minute with a furry tongue. He had never taken the step of having one of those new 'biotronic chips' that multiply the effects of sleep implanted under his scalp. One hour's sleep with these was equivalent to seven hours of 'natural' sleep. All high-ranking imperial officers had undergone this small and extremely practical operation to save time for work and avoid wasting precious hours sleeping. All officers, that is, except Dimitri: the prospect of becoming a 'bionic man' scared him. He was actually disgusted by men of this sort – an increasingly common sight – who suffered from neither heart problems nor diabetes and had artificial ultra-performing hearts or livers implanted. At the age of 68, he was as fit as a fiddle. Now that cancer and cardiovascular diseases had disappeared among the executive elite of the Empire, his life expectancy was 105 years.

The business meeting with the Ministry of the Navy of the autonomous state of Brittany had gone on until 2 AM, so long had

---

[1] This seems to be a play on two famous works of Russian literature: *One Day in the Life of Ivan Denisovitch* by Aleksandr Solzhenitsyn, about a man imprisoned in a Siberian gulag during the Stalinist era, and *Oblomov* by Ivan Gonchorov, about an aristocratic man who refuses to get out of bed.

it taken Dimitri to get those Celts – stubborn as mules – to reach an agreement.

The electro-taxi was waiting outside the hotel. Dimitri said the word 'station' into the microphone of the vehicle's on-board computer, followed by 'fast, arrival at 8:10 AM sharp – I cannot miss the Brest-Moscow-Komsomolsk train' and inserted his credit card. The computer answered in an artificial female voice, 'Brest Urban Transport welcomes you on board pilotless electro-taxi 606. Your request has been processed. You have a 76% chance of reaching your destination on time – the traffic is flowing smoothly. You have been charged 8 Eurosesterces.[2] Please take your card.' Dimitri understood Breton like most of the educated leaders of the Federation. It was a chic and snobby language used in intellectual circles, just like Latvian, neo-Occitan and Basque. The voice repeated the information in Russian, as the credit card suggested this was Dimitri's mother tongue.

The automatic vehicle made an abrupt start. Guided by its electronic maps, it whizzed towards the station. At that hour of the day, the traffic was indeed running smoothly, with only a few carriages, cyclists and knights on the road, and a carriage drawn by a sturdy white horse. After a few sharp swerves, electro-taxi 606 stopped in front of the station run by the TKU (Trans Kontinent Ultrarapid, the bullet-train company). A light drizzle was falling from the sky, which looked heavy, low and grey. The weather was hot and sticky. With climate change, the climate of Brittany had become humid and tropical. Dimitri was impatient to enjoy the icy air and blue sky of Dorbisk, his home on the Bering Strait, 20,000 kilometres away, at the other end of the vast Eurosiberian Federation – the 'Great Homeland.'

8:17 AM

The train silently left the underground station. Dimitri Leonidovich immediately felt the effects of its powerful acceleration. On the screen embedded into the back of the seat in front of him he studied the schedule and route of his journey: Brest-Paris-Brussels-Frankfurt-Berlin-Warsaw-Kiev-Moscow… down to Komsomolsk, on the banks of the Amour River, in the Siberian Far East. There he was going to catch a plane straight to Dorbisk, as the track for the planetrain to the Bering Strait had not yet been completed. Dimitri was going to spend

---

2   The *sesterce* was the coin of the Roman Empire.

A DAY IN THE LIFE OF DIMITRI LEONIDOVICH OBLOMOV

the night with his wife Olivia to celebrate their ten-year anniversary. In Brest it would be just past 3 PM, but in Dorbisk, because of the time difference, it would be 2 AM...

All this was possible thanks to the planetrain or 'planetary train', as it was officially called. This revolutionary invention had radically changed the world of transport just after 2040. The patent for it was an old one: it had been registered by the (now defunct) American company Westinghouse in 1975! The principle on which it was based was the following: along a tunnel dug a few metres beneath the earth, a train – or rather a semi-articulated train of 150 metres in length functioning through magnetic levitation and 'electro-linear' propulsion – runs in a vacuum-packed atmosphere. Given the absence of friction from either the air or the ground, the planetrain can travel as fast as 20,000 kilometres per hour. It cannot travel at its full speed over short distances because of acceleration and deceleration problems, reaching 1,300 kilometres per hour at most. On long distances, however, it attains close to 20,000 kilometres per hour. Hence, the journey from Brest to Paris (480 kilometres) took longer than going from Moscow to Irkutsk (7,000 kilometres), as in the latter case the train could reach up to 17,000 kilometres per hour, albeit only for short stretches along its course. On the whole, the planetrain journey from the Atlantic coast to the Pacific took just over three hours.

Following the traumatic occurrence of the Great Catastrophe of 2014-2016, the 'Renaissance' of 2030 and the building of the Eurosiberian Federation, which was given the name of 'Empire of the Two-Headed Eagle' – for it marked the fusion between the European Union and Russia with the Pact of Prague in 2038 – the revolutionary Federal Government had chosen to make a clean break from the ideas of the past in the field of transport, as in all other fields. The use of electric vehicles had been extended to all, while private car ownership had been halted; horse power as a means of transport had returned, while the use of engine-driven vehicles in neo-traditional rural communities had been banned; highways had been abandoned and replaced with railway tracks for fast trains carrying lorries and containers; air travel had gradually been phased out in favour of planetrains; cargo-airships had been introduced for shipping goods; the canal network had been restored; and, finally, nuclear energy was being employed along with wind energy for maritime transport. The Government had been imposing these radical changes – a clean break from the past – since the '40s, and this had been possible because it was necessary to start from scratch. Once destroyed or rendered unserviceable by the Great Catastrophe, economic systems and infrastructures had been rebuilt on completely new foundations.

The construction of the planetrain, like other great continental projects, had enabled the launching of a new techno-scientific economy between 2040 and 2073. Unlike in the Twentieth century, this was no longer extended to all areas of the Earth or to all people: only 10% of humanity benefited from it. These people were grouped in cities – far smaller and less densely populated than Twentieth century ones. Within the Federation, 20% of the population lived in techno-scientific industrial areas. This had made it possible to repopulate deserted rural areas and solve the problems of pollution and energy waste – the planet could finally breathe again. The biggest city of the Federation, Berlin, only had 2 million inhabitants. Still, it was too late to stop global warming, the greenhouse effect and the rise of sea levels caused by wide-scale toxic emissions in the Twentieth century. Science had made rapid progress, but it only affected a minority of the population; the others had reverted to a Medieval form of economy based on agriculture, craftsmanship and farming.

The reason for this dynamism is that the global volume of investments and budgets, both public and private, no longer had to meet the various needs of 80% of the population, who now lived in neo-traditional communities based on archaic socio-economic systems and personally managed their own production and exchange of goods. So starting around 2040, innovation in technological science had resumed the level it had reached in 2014, but only in certain spheres: transport, computer science, genetics, energy, space exploration, etc. In all other sectors, given the limits of the market, technological products were rather primitive. Basically, a two-tier economy had been established.

Seven planetrain lines had been built between 2040 and 2073, all of them connected: Brest-Moscow-Dorbisk, Rome-Edinburgh, Lisbon-Oslo and St. Petersburg-Athens were already finished, while others – such as the Helsinki-Vladivostok line – were still under construction. Outside the Empire, only China (Peking-Shanghai) and India (New Delhi-Bombay) had bought plaintrains, which were jointly produced by the Typhoone and Eurospace companies. America, which had never really recovered from the Great Catastrophe and had almost entirely reverted to an agrarian economy, could not afford to pay for them. Besides, long-distance connections down there only interested a very few people: for only 8% of the American population lived in a techno-scientific system, chiefly along the Pacific coast and around Chicago. Even air travel was rare and made mostly via airships, since – after the Great Catastrophe and the devastating consequences of the greenhouse effect – a phobia of jet planes had spread. The days in which people – like Dimitri Leonidovich's great-grandparents – dreamed of supersonic jets were truly dead and gone...

Brest-Berlin

The screen in front of Dimitri displayed the speed of the underground train: 1,670 kilometres per hour. On a simple map, a luminous dot indicated its position: ten minutes away from Paris Montparnasse. Paris... A city that must have been magnificent in the Twentieth century, Dimitri thought. He had few memories of it. He was only ten in 2016, when his family had fled the city plagued by anarchy and hunger to return to Russia. Most of the monuments had been burnt and destroyed, and its museums and treasures had been pillaged during the civil war that had broken out before the Great Catastrophe. Today, the autonomous state of Ile de France was carrying out restorations and reconstructions, but Paris was unlikely to ever return to its former glory. The only way to learn what the *Mona Lisa*, Sainte-Chapelle, the Eiffel Tower or the Louvre looked like was to visit virtual Websites with 3D images.

Dimitri Leonidovich sighed in sadness at these unpleasant thoughts and took out his multi-purpose laptop computer – every high-ranking imperial officer had one –from its case. This was a genuine wolf-fur case decorated with a double-headed Eagle on a red-and-white chequered background.

Dimitri opened the small object, which served almost any purpose. He adjusted the screen and keyboard and immediately Vega, his 'virtual secretary', appeared in 3D. He had created an ideal female assistant for himself on his quantum computer to be the opposite of Mrs. Groux, the dreadful and all too real secretary who worked for him in the headquarters of the Imperial Government in Brussels – a fat and repulsive old hag. His virtual secretary Vega had perfect measurements, always appeared in scanty dresses and made suggestive remarks from time to time; she knew all of Dimitri's life and shared his intellectual outlook.

Named after one of the stars shining in the Siberian sky, she was the woman of Dimitri's dreams. He had created her in secret, keeping her existence concealed from his wife Olivia, who did not know the access code to the programmes of this extraordinary GPT (Giga-Power of Treatment) quantum computer which the huge Typhoone company had produced exclusively for the new aristocracy: the upper echelons and high-ranking civil and military engineers of the Federation. The GPT also served as a mobile phone, a fax and a multi-purpose terminal connected to the Euronet, and could communicate with the whole world by satellite, even from inside railway tunnels.

To prevent those next to him from overhearing his conversation (the planetrain travelled in the vacuum, magnetically suspended, and thus made no noise whatsoever), Dimitri put on his earphones. He switched the machine on and then typed 'Vega'.

The first words of his virtual secretary were, 'I went for an evening dress. It's black and see-through. Do you like it, Master?' A luscious and curvy brunette with a mischievous nose and sultry look, Vega had been meticulously designed by Dimitri with the help of a VSP (Virtual Service Personnel) programme. She slunk sensuously across the small screen in 3D.

Dimitri replied, 'That's perfect, Vega. I am now on the bullet-train, returning from an arbitration meeting in Brest. I will be spending fifteen days' holiday at home, in eastern Siberia, before visiting Brussels again.'

The beautiful girl smiled and stroked her hips.

'Master, I suggest you disconnect from the small screen of the GPT computer and plug into the one in the seat in front of you. You'll be able to see me in a larger format.'

Dimitri hadn't thought of that. He unrolled a tiny wire which he plugged into the screen embedded in the seat. Immediately the image of the virtual girl appeared in a larger size. She continued, 'I would like to remind you that today is your wedding anniversary. You should get your wife a present.'

'I have.'

Dimitri was bringing his wife a Celtic jewel in solid silver from the autonomous state of Brittany: a cross inscribed within a solar wheel with interlaced motifs and a large ruby in its centre. He had found it in a crafts market in the rural community of Landéda, near Brest.

'I disconnected my private phone. Has anyone called?'

A DAY IN THE LIFE OF DIMITRI LEONIDOVICH OBLOMOV

'You have received two messages. Would you like to hear them?'

The first message was from Olivia, who confirmed she would be waiting for him at the airship port in Dorbisk.

The second message was from his friend Hans Gudrün, the governor of the state of Bavaria and a member of the Central Committee of the Federation (the body representing autonomous regions before the Imperial Government).

The Bavarian had called him on his videophone. An icon appeared in the top-left corner of the screen showing the smiling, ruddy face of the governor, who was wearing a green, feathered hat.

'I hope you managed to solve our problems with those stubborn Bretons and defend Bavaria's point-of-view. Expect a far more difficult negotiation after your holidays. Bavaria disagrees with the federal project for a solar powered high-energy plant. I hope you will take our view into account, my friend. Send my regards to Olivia and your children. I have booked a place of honour for you at the Munich Bierfest in September. *Tschüss!*'[3]

Dimitri would have to phone him back later. Gudrün was very kind, but shouldn't put pressure on him like that, using their friendship as an excuse.

'Any other news, Vega?'

'Yes, Master. The last EKIS bulletin contains information that might interest you.'

The EKIS, or 'Euro-Kotinent Information Service', was an information network exclusively reserved for the leaders and cadres of the Federation. The media system that had been open to all in the Twentieth century had gradually disappeared, for it was thought to cause disinformation and demoralise the public by causing panic. With the help of keywords, Vega had selected news of interest for Dimitri.

'I'm listening.'

The image of the virtual secretary shrunk to the size of an icon as a voice-over commented on the images now flashing across the screen. Vega had selected many news items, according to the centres of interest programmed by the Councillor. Dimitri focused his attention and fastened his seatbelt, for the train was rapidly decelerating and entering the underground station in Paris.

---

3   German: 'bye'.

'Demonstration outside St. Peter's in Rome for the return of the Pope' (a crowd was shown holding placards outside St. Peter's, which was covered in scaffolding. The Roman Republic was rebuilding the Basilica, which had been destroyed during the war against the Muslims). The voice-over continued:

'As is widely known, since the murder of the last Pope, John Mary I, in 2017, and the Great Catastrophe, no Pope has been elected. The Holy See collectively administers the Church. Since the schism of 2020, with the election of Popes Pius XIII, Pius XIV and now Pius XV, who is residing in Avignon, the traditionalist Church – which has been declared "impious" – has been calling for the return of "its" pope to Rome and the Vatican. The Holy See is refusing to meet this request, leading to the present traditionalist demonstration. Some protesters have travelled all the way from Poland by cart – a three months' journey. No accidents have been reported so far. The Senate of the Republic of Rome is backing the Holy See and opposing the return of the Popes, in compliance with the Concordat of 2022 and in agreement with Father Diaz Fernandez, Superior of the Society of Jesus (the Jesuits). The Imperial Government has issued a bulletin stating that, in conformity with the principle of religious neutrality, it will take no side in the conflict, for this concerns an authorised religion within the Federation, Christianity. The druidic representatives of the Great Brotherhood of Cernunnos,[4] assembled in a conclave in London and representing all Celtic pagan cults, have issued a statement calling traditionalist Catholics 'to join them'. The Imperial Government and the Central Committee of the Party remind all civilian cadres and members of the armed forces that they must not become involved in these disputes and must keep a strictly neutral stance.'

The demonstration outside the Vatican disappeared from the screen and was replaced by the image of a knight in armour from Poland, waving a white flag emblazoned with the monogram of Christ amidst the applause of the crowd.

After a 'beep' a new image appeared. In a hangar, a bizarre engine with enormous solar panels was shown, as big as a railway coach and surrounded by engineers at work. The voice-over explained, 'This is IPC, the new Ionic Propulsion Cruiser developed by the Typhoone

---

[4] Cernunnos is a god of the Celtic religion. Icons of him have been found in France and Germany.

A DAY IN THE LIFE OF DIMITRI LEONIDOVICH OBLOMOV

company and by Euromotor on the basis of a 1995 project[5] that was lost and has now been rediscovered. More efficient than space vehicles with conventional propulsion, the IPC can reach our base on Mars in two rather than nine months thanks to its "gradual acceleration" from the orbit of the Moon. It is fuelled by xenon, a rare, electrically charged gas that can be easily stored and which sets off a flux of high-energy ions. This sequence of images was filmed in an IPC assemblage factory in Toulouse, in the Occitan Republic.'

This was followed by another scene: a huge missile bearing the red-and-white chequered flag of the Empire was launched with a pyrotechnic show of lights and smoke. The voice explained, 'Yesterday, at 2:45 AM GMT, the first IPC with five astronauts onboard was sent into lunar orbit by a Leonida missile, which left our floating platform for equatorial launches in the middle of the Atlantic. This revolutionary spacecraft will reach our base on Mars in 60 days. We are now well ahead of the Chinese and have a decisive advantage over them for the conquering of Mars.'

The image of the missile, whose white banner disappeared above the clouds, was replaced by a gaily coloured feast: bare-chested men, girls dancing with embroidered dresses, beef roasting on embers...a merry crowd of farmers. This was taking place at the centre of a vast clearing. The camera moved across the landscape: mountain peaks dotted with tight rows of white villages. The voice commented, 'This is the feast of the summer solstice in the Republic of Lacedaemonia, which includes the Peloponnese. Since 2030 we have been witnessing a huge renaissance of this ancestral custom and it now represents a key moment in the life of many rural communities of the Federation. On the night of 21 June, the longest of the year, a large pyre is lit [the image of a brazier was shown]. For three days, huge festivals take place. Farmers, sailors, craftsmen, as well as engineers and imperial officials assemble from the four corners of the Empire to take part in this folk celebration in the ancient city of Sparta that stretches back into the mists of time.'

This was followed by some interviews: one with a Provencal mutton farmer who had travelled forty days by horse to reach Sparta ('My sheep are well protected against wolves: I have three daughters and two guard dogs'), and another with a Swedish cosmonaut and Odinist who

---

5 This probably refers to Deep Space One, an ion-propelled space probe that was launched as an experiment by NASA in 1998. The contract was secured by Hughes Electron Dynamics in 1995.

had arrived with his wife and six children on the Northern Europe-Athens bullet train and then a mini-airship taxi ('We live near local people, in a rustic house, and wash with water from wells – but it's still a good deal more comfortable than the Moon base!').

The commentator, most probably a member of the Party, ended his report with the words, 'All members of the Federation should bear in mind that the Sparta Solstice celebration is entirely self-funded.'

Berlin-Warsaw-Kiev

The planetrain came to a halt in the underground station in Berlin. Time – and the stops in Paris, Brussels and Frankfurt – had passed unnoticed by Dimitri. During each acceleration and deceleration he had mechanically fastened and unfastened his belt.

A flock of screaming, playful children swarmed into the compartment. From their uniforms one could tell this was a group of 'Eaglet' scouts, the youngest division of the federal youth organization. They were over-excited at the prospect of boarding a bullet-train for the first time. They were no doubt going to attend a camp in some forest in the Urals or Siberia. These camps were very popular.

One of the kids accidentally hit Dimitri in the face with his backpack. The leader of the group – a Valkyrie with an exquisite body – apologised profusely (seeing Dimitri's prestigious Plenipotentiary Councillor's uniform). She shouted in German at the kids, who suddenly became silent and took their seats.

Following the Renaissance of 2030, the demographic winter and the depopulation caused by the Great Catastrophe, demographic levels had risen again, as if the collective biological unconscious had been awakened. Now children were everywhere. Losses needed to be made up for, although 18% of the births among members of the elite were assisted by genetic engineering: pregnancies in incubators – saving women the trouble – ensuring a 'planned genome improvement'. Use of this technology, however, was strictly banned in the neo-traditional communities and in any case subject to the approval of the Imperial Eugenics Committee. Children born through artificial procreation were often consecrated as 'wards of the Empire' and assigned to educational centres where they were trained to become ultra-performing cadres. China, the Federation's great rival, had also adopted this policy; in the field of eugenics, it even held a certain advantage.

The train decelerated again. It was now reaching Warsaw. A dark-skinned and very beautiful girl with long, jet-black hair down to her shoulders and dressed in a violet sari stopped in front of the empty seat next to Dimitri.

'I haven't made a reservation, but can I sit here?' she asked in English, pointing to the empty seat.

'Please do, Miss...'

Dimitri's heart rate increased slightly. The foreign girl gave off a sweet scent. As was customary, she introduced herself with an enticing smile.

'My name is Nafissa Godjab. I am the daughter of the Maharaja[6] of Gopal, the Indian Minister of Foreign Affairs. I have just completed a two months' study programme in the Eurosiberian Federation.'

Dimitri in turn introduced himself, specifying his rank. 'I'm the Plenipotentiary Councillor of the Inter-State Court of St. Petersburg, to which I answer. My role is to resolve conflicts within the Federation. I am also responsible to the Imperial Government in Brussels, where my offices are located. I am now returning from a meeting that took place in one of our states, Brittany, and will be joining my family for a ten days' holiday in my native town, Dorbisk, in eastern Siberia, on the shores of the Bering Strait.'

The Indian girl gazed at Dimitri's uniform with a silent smile.

'So you're an important man, then? And no doubt a very cultured one, too?'

Dimitri wasn't sure what to answer. The young aristocrat was making a strong impression on him and he could feel himself blush. He said, 'I have a daughter your age. Her name is Lizia. She looks like you, although she's blonde; she's as charming as you are. She's studying history... As for whether I'm an "important man", this is a different matter. I serve the Great Homeland and travel across it, far and wide, to ensure its unity...'

The girl didn't answer. She lowered her eyes and took a small recorder out of her tiger-skin bag.

'Mr. Councillor, in the Indian Empire they don't teach world history very well. It is as if they wish to hide what happened. Not even my own father will speak a word about it. What happened after the end of the Twentieth century? In my country people speak of a "Great Rapture".'

---

6   *Maharaja* is Sanskrit for King.

Nafissa was speaking in a low voice, staring at Dimitri with her wide, black eyes. The Councillor couldn't refuse any request from a daughter of the Minister of Foreign Affairs of the Indian Empire on a study exchange in the Federation. It was a diplomatic requirement that he answer. And besides, she was so pretty... So Dimitri decided to hold a short history course.

Acceleration pinned them to their seats. The screen in front of them displayed the speed of the train: '7,800 kilometres per hour. Next stop Kiev, in 15 minutes.' A list of airship connections for a dozen Ukrainian cities followed.

'The world you know today,' Dimitri started explaining to the girl, 'has little to do with that of the Twentieth century. The civilisation that had developed between the Sixteenth and Twentieth centuries and had progressively spread globally – the period which reactionary, backward-looking idiots continue to refer to as the Golden Age or "500 years of glory", and which they would like to restore – was founded on utopia and ended with a deadlock and monstrous collapse. In line with the scientific predictions made in the late Twentieth century, and which governments ignored, this global civilisation and politico-economic system brutally plunged into chaos because of a dramatic convergence of disasters of all sorts that multiplied, according to the "chaos theory" or "catastrophe theory" developed by Twentieth century mathematicians René Thom and Ilya Progogine.'

'How did this happen? Have you got any memories of it?'

'I was ten when it all happened. The explosion hit suddenly, without any warning, in 2014. But of course you are so young...'

Dimitri gazed intently into the eyes of this 20-year-old Indian beauty. His gaze then lingered, almost involuntarily, on the girl's breast, which was protruding from under her sari.

'Please answer my question, Mr. Councillor, and stop giving me the eye – it's not proper. I should remind you that in the Indian Empire, interracial love affairs are strictly punished, even when they take place abroad.'

Nafissa was speaking calmly, with a smile on her face. Dimitri blushed and cleared his throat.

'But that wasn't at all my intention. Now, let me answer your question. First off, "chaos theory": any system, be it a civilisation, moving vehicle, drop of water on the wing of a plane, climate condition, human relationship or living being, is a form of balance deriving from complex interrelations. It is enough for a single parameter to change for the whole system to suddenly fall out of balance: the civilisation will crumble, the drop of water will fall off the wing of the plane, a storm will break out, a couple will divorce, symptoms of illness will appear, and so on. The system, in other words, will disappear – this is the "catastrophe". Then, after a period of latency and resetting – the "chaos" – a new system will come to light, one based on different relationships. This is precisely what happened to the global civilisation of the Twentieth century. It was too big a bubble not to burst.'

'I think I understand. But how did it all happen? I'm interested in this because I also study traditional theatre and would like to write a piece about this mysterious "Great Catastrophe".'

'What?' Dimitri said with surprise. 'Don't they teach you anything in Indian schools? Have you never studied history?'

'No. In my country, the people in charge have decided to remain silent on this matter. They pretend nothing happened – that the "old world" never existed. No doubt, because they fear people may want to restore this ancient civilisation and return to the Western model. Besides, we don't have any "history courses". The word itself doesn't exist. History for us doesn't exist: what we are taught about are our ancestral traditions and the lives of our gods. Of course, I belong to the caste of those who have preserved a technological lifestyle and have a passport to travel abroad, but still...'

'But what?' said Dimitri, who was impressed by the intellectual brightness of the Indian girl.

'Your "theory of catastrophe" is simply what our poets call the mechanism of tragedy. As I told you, I'm studying theatre. The ancient Greeks, too, used to say the same thing.'

Kiev-Moscow

A beep was followed by a blinking red light. The screen announced, 'Fasten your seatbelt. Deceleration level G2.[7] We are arriving in Kiev.'

Dimitri went on, 'It was the year 2014 and my parents were working as Russian diplomats in Paris. I was ten years old at the time and was attending an international school in the 16th arrondissement, near

---

7   G, which is short for gravitational force, refers to the amount of gravitational force acting on a body when it is accelerating. The amount of force acting on a stable body on the Earth's surface is 1G. A vehicle which accelerates rapidly, such as a fighter jet or the planetrain in this story, would subject its passengers to a high level of force – in this case, twice what a person ordinarily experiences in everyday life.

the embassy. I can remember it as if it had happened yesterday. I was very mature for my age. That year, 2014, really was a black one. It took us by storm: tragedy, as you say, occurred all too suddenly.'

The Russian councillor was speaking in a low voice, broken by emotion. Clearly, he was reliving a traumatic moment in his life. Only the charm of beautiful Nafissa persuaded him to continue his narrative.

'Were there any signs of warning?'

'Yes. Symptoms of the tragedy were already becoming clearly visible, according to historians, in the late 1970s and then became even clearer in the 1990s. According to the chaos or catastrophe theory outlined by Thom and Prigogine, the changing of a single parameter is enough to make a system collapse. This is the so-called "butterfly effect". In this case, a dozen changed parameters were converging!'

The girl was hanging on Dimitri's every word.

'So, how did it all start?' she whispered.

The train came to a halt in the underground station of the capital of Ukraine. Some people got off, while others filled what seats still remained vacant. Dimitri noticed the presence of several imperial and military officials, wearing a dark-violet uniform with a golden shark on its collar badge. These were officers from the H.L. – the 'Hoplite[8] Legion': the elite troops of the Federation.

As the planetrain set off, they were again pinned to their seats. On the screen a sign in various languages read, 'We are currently travelling at a speed of 14,000 kilometres per hour and will be reaching Moscow in 10 minutes.'

Dimitri went on, 'Ethnic revolts had been breaking out in Paris and other big European cities for a number of years. No government had managed to curb unemployment. A year of slight improvement was followed by an even more serious decline. Poverty spread and it became practically impossible to leave one's house after sunset. The ageing of the population had destroyed the social security and pensions system, and the flight of intellectuals and unchecked immigrations made things even worse. Gangs of thugs and descendants of immigrant families led to a climate of unbearable insecurity in the cities, including in neighbourhoods that had previously been spared this plight. A sort of rampant and endemic civil war had broken out,

---

8   Hoplites were the soldiers of ancient Greece.

A DAY IN THE LIFE OF DIMITRI LEONIDOVICH OBLOMOV

which the police could hardly control. Starting in 1998, particularly in France, ethnic gangs from the *banlieues* made a habit of regularly pillaging and looting town centres.'

'But why didn't people and governments react?'

'They were paralysed by a jumble of old humanitarian ideologies. And besides, after the Amsterdam Treaty of 1999,[9] not only did individual European governments have hardly any real power, but even the embryonic European federal government still didn't have any. This interregnum was a time of paralysis. To cut a long story short, between 1999 and 2014, the year of the global explosion, France dragged Western Europe into the abyss. Everything came together and added up with increasing effect: the economic crisis, impoverishment, latent ethnic conflict... Starting in 2002, the gross domestic product of Europe dwindled and then hit rock bottom.'

Nafissa continued focusing on the Councillor's words.

'You are contributing to my thesis with some extremely interesting facts. We know nothing about these things in India.'

The girl drank a glass of 'Regenerator' served by the hostess. This was a vitamin-rich drink with slightly euphoric effects which was perfectly harmless but was completely unavailable to ordinary people. Dimitri continued staring longingly at Nafissa.

'In fifteen days I'll be back in Brussels. Come visit my office: I can give you many documents on this historical period in support of your thesis... I would also like to take the chance to invite you for dinner in an excellent tavern run by some monks.'

'Does "Vitalist Constructivism" authorise you to do so?'

Vitalist constructivism was the official ideology of the Federation.

'Considering your rank, I don't think this will be a problem. You must have an international alpha level certificate, right?'

'Yes, thanks to my father. I have the right to come and go wherever I like in your Empire.'

With a smile, she took out a plastic-coated gold card adorned by a white dove with a red key in its mouth: the pass the Federation issued for foreigners. Nafissa burst out laughing. She then stopped and asked:

'Did no one resist? Why did the state give in? Why didn't people react? I'm talking about France, the place where you say it all started...'

---

9   The Amsterdam Treaty, signed in 1997 but which went into effect in May 1999, in part gave more powers to the European Parliament of the EU.

'Well, yes: some people did react. There was a political party, the Front National. They had been seeking to prevent the catastrophe since the 1980s. But theirs was an impossible task. The party was demonised by the elites – deeply masochistic elites, which collaborated with the enemy. A dying folk is always fascinated by the abyss. The Front National tried to react, but in vain. In 2014, it received 30% of the votes in France despite the increasing number of descendants of immigrants and newly arrived immigrants from the southern countries.'

'In India there is a saying that goes: "It is never men who do things, only Shiva."'[10]

Moscow Station

The car started shaking slightly. It slowed and entered the underground station in Moscow. Dimitri explained, 'The atmospheric pressure is normalising. The bullet-train is shaking because air molecules are hitting its cockpit. Don't be afraid.'

'I'm not afraid. In India they also teach us some physics....'

'Your Indian proverb is most apt. Humans have no wisdom: they always do things at the very last moment. People only react when a cataclysm hits them – which usually means when it's too late, as was the case here. Instead of carrying out reasonable reforms before the tragedy occurs, they prefer to carry out brutal, terrible revolutions later. This is exactly what happened. It is God who forced us to reset out clocks. It is He who governs our destiny.'

'No. It's the gods,' Nafissa said in a low voice.

'The sinister year 2014 witnessed the convergence of four events: in France, revolts of an unprecedented level of violence broke out; the police were overwhelmed and the powerless government did not dare call in the army. That year, the endemic uprisings caused by the (usually armed) ethnic gangs that moved from their lawless enclaves to attack city centres turned into a real insurrection, which ravaged France between 2014 and 2016. The political elections of February 2014 only brought things to a head. An increasing number of voters were of immigrant origin, and so what had been predicted in the 1980s finally happened: the Parti Populaire Musulman (Popular Muslim Party or PPM) received 26% of the votes and the Front National 30%.

---

10 Shiva, one of the major deities of the Vedic pantheon in Hinduism, is the god who destroys the universe at the end of each cycle of time. Hindus who elevate Shiva above the other gods are known as Shaivites.

A DAY IN THE LIFE OF DIMITRI LEONIDOVICH OBLOMOV 211

Things quickly escalated from there. The "secular and Republican" centre-Left coalition was no longer able to govern. The demands of the PPM became increasingly unacceptable. Some people accused them of wanting to turn France into an "Islamic Republic". One of the party's extremist leaders replied: "Yes, for within ten years we well be in the majority. By then, France will be an Islamic land. This is our revenge for the Crusades and colonisation!" The Front National then issued a call for "Resistance, Reconquest and Liberation". It is in this context that the extremist Muslim leader of the PPM group in the National Assembly was murdered.'

'By a member of the National Front, I guess?' Nafissa asked.

'No. Probably by the Algerian secret services, in order to spark a revolt among Muslims in France. Bear in mind that since 2004, North African countries had turned into fundamentalist Islamic Republics that were extremely hostile to France. In other words, this murder signalled the beginning of a widespread revolt of an unprecedented level of violence.'

The girl gave Dimitri a wide-eyed look of astonishment.

He went on, 'In a short time, the plague spread to England and then Belgium and Holland – countries which also hosted large immigrant communities and where Islamic parties similar to the PPM had many voters and an ambition to seize power. The European government in Brussels was utterly at a loss. That's when the first wide-scale strikes took place. The economy was gradually paralysed and then shortages of basic goods like water and food began. My family stayed in the embassy with other diplomats. We didn't dare go out. The rioters were setting fire to buildings in the town centre and the streets echoed with gunshots. Still, no order was given to the army to intervene! The police were overwhelmed. The Front National set up "patriotic self-defence militias" and a "National Resistance Council". But it was too late: the French Republic, civil order and the economic system were all collapsing. Gradually, people fled the cities. A terrible economic crisis followed the civil war.'

'Did no one manage to re-establish order?' the Indian girl asked in amazement.

'No. Ours was a society that was growing old and was undermined by the viruses of pacifism and humanitarianism. It was incapable of defending itself. Consider that between 2014 and 2016, part of Western Europe – France, Great Britain, Belgium and Holland – quite simply returned to the Middle Ages. Even international aid could not reach us because of the civil war. It is now believed that 40% of the population in this area died as a result of war, famines and epidemics! In only three years, part of Western Europe plunged into anarchy. States simply disappeared. The government in Brussels was no longer of any use. Armed gangs scoured the countryside in search of food. Trains and cars stopped running. The French fled to refugee camps in Germany, Italy and Spain. And there were no longer any television broadcasts...'

'Any what?'

'Television broadcasts. The television was an old on-screen broadcasting system whereby the whole world could view the same images at the same time. It had turned into a kind of religion or drug. But let's move on, this is just a thing of the past...'

After suddenly leaving the Central Kremlin Moscow station, the bullet-train took up speed. 'Further on, towards the Urals, lies my homeland – Siberia,' Dimitri thought to himself. He imagined the train, like a cobra, dashing towards its prey... Again they were pinned to their seats, like Tintin,[11] pressed against his bunk by the formidable acceleration of an atomic-powered rocket in his journey to the Moon. Tintin – that old comic character from the Twentieth century, whom only men of letters knew now...

Moscow-Yekaterinburg

'And what about you and your family? Did you return to Russia?'

'Yes, along with all the members of the embassy. We were repatriated in a rather extraordinary way two months after the revolt had broken out. Things in Russia weren't that great, but compared to France it was paradise! After the fall of Communism in 1991, the new regime proved incapable of converting to the free market economy. The country was collapsing. Then, in 2002, a nationalist and neo-Communist military regime seized power. Since 2014, something close to a dictatorship had been installed: Russia was autarchic, but still – despite widespread poverty and the collapse of the capitalist dream – there was enough food for everyone. So I resumed school in my own country. Russia, which in the year 2000 had been the sick man of Europe, fourteen years later, in the midst of all the chaos, was just about the

---

11   Tintin is a character who travels the world in *The Adventures of Tintin*, a series of comics which began in Belgium in 1929.

only country in which civilisation hadn't collapsed and one could find a degree of safety and order.'

'There is one thing I do not understand.'

The girl's dark-green eyes met Dimitri's.

'How could the collapse of the countries of Western Europe, which made up only a small percentage of the world population, cause what has been called the "Great Catastrophe"?'

'By an avalanche effect. According to the mathematical catastrophe or chaos theory, for a stable system to topple over, it is not necessary for most of its elements to disintegrate. All that is needed is to change a central parameter. Now, the Western portion of the European continent was a main parameter for the balance of world civilisation and its economy. Besides, as I already mentioned to you, what occurred was a convergence of various other "mini-catastrophes" that affected the planet but had already been quite foreseeable by the 1980s. Starting in 2015, the Mediterranean and central Europe, including Germany, experienced the same tragic events as France, England, Belgium and Holland, and to their full effect.'

Dimitri searched in the girl's eyes for the impact of his words and only found great curiosity. 'She really has an enchanting gaze,' Dimitri said to himself. He focused for a moment on the image of Olivia, who would be waiting for him in Dorbisk that night. He then continued his narrative, 'The European economy as a whole collapsed like a deck of cards. Between April and December 2014, a civilisation disappeared, just like that.'

'And what were the consequences for the rest of the world?'

'The events taking place in Europe, which had been the greatest economic power in the world, caused a recession such as had never been seen before. In June 2015, the President of the IMF[12] uttered words that are now part of history: "This is not an economic crisis. This is not a recession. This is the end of the modern world: this is the apocalypse."'

The Indian girl smiled. 'That was the gods' will.'

She added, 'And what were the other three tragic events of the year 2014?'

'The first was a global financial crisis, similar to the one that had occurred in 1998, only a hundred times worse. This crisis coincided

---

12 The International Money Fund, an international organisation intended to help stabilise the global economy.

with the outbreak of civil war in France. So there was a cumulative effect. The world economy, which had grown weak because of its financial and speculative foundations, popped like a balloon. The second event was a nuclear war between India, your country, and Pakistan. It is as a consequence of this that you have annexed Pakistan and recreated a unified subcontinent like the one which existed under British colonial rule.'

'This I know, but Pakistan attacked us!'

'In war, no one is simply the attacker or the attacked: one is both things at once. Relatively speaking, this war didn't cause a huge number of deaths – two million at most – but it was a global shock that destabilised the system. It was China, which threatened to intervene, that brought an end to the conflict and authorised the annexation of Pakistan, following a bizarre plan and despite its historic enmity towards India. The United States could not prevent this from happening. What had been a leading world power that had dominated the Twentieth century disappeared like a comet, as quickly as it had emerged.'

'The United States was the name of North America, right? Today it is almost impossible to imagine that this region dominated the planet in the late Twentieth century...'

'Indeed. History is unpredictable: it is written by blind madmen and sleepwalkers. The same thing had happened to the Spanish empire, a long time before.'

'And what was the third event?'

'An environmental catastrophe similar to what humanity had already experienced in the 1990s, only this time on a far wider scale. In January 2014, millions of hectares of rainforest caught fire in the Amazon – from deforestation work by large agricultural companies. The Amazon, the green lungs of the planet, lost 30% of its surface in one year (which is as long as the fire lasted). The smoke and dust that filled the atmosphere blocked the rays of the sun for six months, causing major climatic disasters across the world: devastating cyclones, torrential rains and droughts which further contributed to the environmental damage that had already long since begun, in a variety of ways. The psychological impact of all this was huge. To make things even worse, the ocean level rose as a consequence of the greenhouse effect: the use of greenhouse gases since the onset of the industrial revolution ultimately led to global warming and the melting of the ice caps. In September 2015, with the equinox tide, a huge wave hit the Atlantic coast. In the centre of New York the water reached two metres in height and coastal cities in Europe were devastated... All these events added up, with consequences on both a physical and psychological level. For the whole world, the years 2014-2016 were a great upheaval. The civilisation of "modernity" disappeared in three tragic years to make way for a different world.'

Yekaterinburg-Novosibirsk

The train left the underground station of Yekaterinburg. After that 2,000 kilometre leap, cruising speed reached 12,000 kilometres per hour in just a few minutes – about half the speed of the orbital station Leonardo da Vinci. Dimitri pictured the taiga a few metres above their heads, crossed by packs of wolves and by the heavy wagons of lumberjacks making their way back along some path from their clearing areas.

'Please continue your narrative, Mr. Councillor. I am learning a lot of history from you.'

'Things took place – or rather exploded – between 2014 and 2016. It was like the collapse of the Roman Empire, only on a vaster scale and with an acceleration of history. By 2016, the area that included France, Great Britain, Belgium and Holland had plunged into complete chaos: 40% of the population had died as a consequence of the massacres of the civil war, of famines, epidemics, and the collapse of an extremely fragile technological civilisation and global economy. There were no longer any states and the cities were empty. In the rest of Europe frontiers were strengthened to avoid the incursion of armed gangs or refugees. The inevitable consequence of this was that the "global system" crumbled. These events all occurred with frightening speed, spreading like a cancer causing widespread metastasis in a living organism.'

'I heard that there was a Muslim invasion of Europe? Is this a fabrication or is it true? As for us, in India we've completely solved the Islamic problem...'

'In 2017, the Islamic republics of North Africa, which had been established following the 2003 revolution, took advantage of the complete chaos that reigned in France. An invading army landed in Provence and occupied it militarily. It tried to set up an "Islamic Republic of France" and banded together the armed ethnic gangs that scoured the country and fought each other, but failed because of the widespread chaos. It was a new Middle Ages: a return to the Sixth century, with pockets of resistance in various areas setting themselves up as new baronies. The most powerful one was centred in Brussels, the old capital of the European Union. Here, in 2018, the "Duchy of Brussels" was established by a member of the Belgian Army, which had managed to protect the city and free it from the "ethnic gangs", as they were called at the time.'

The Indian girl asked in disbelief:, 'But why didn't the armies of these countries intervene?'

'This is a good question. The reason is that the governments of these countries, which were guilt-ridden and filled with fear, gave orders too late – in the early months of 2017. By then, the economy had collapsed: there was no electricity and no fuel, and the army was paralysed. In fact, there no longer *was* an army. As had happened in Russia twenty years earlier, soldiers were no longer being paid and so were deserting en masse. Only certain areas were protected by officers who managed to restore some order, defeat the armed gangs and ensure supplies in their cities through the control of the surrounding countryside. By the use of force, they also managed to reopen some power and purification plants. Clearly, the regimes of these duchies, which were hardly connected to one another, were of a highly authoritarian, military sort. Still, they ensured safety and bread for the people, and that was enough. These "baronies" housed 20% of the population, exclusively comprised of native Europeans. Clearly, the standard of living in these places had returned to be that of – say – the Seventeenth century. All forms of modern medicine, for instance, had vanished, as there were no drugs available.'

'Where were these "baronies"?'

'There were only a dozen in Western Europe: the Duchy of Brussels, the Republic of Brittany – the largest of all, governed by officials of the old French war navy – and various other small ones, centred around Western European cities. They kept in touch with one another by radio.'

Nafissa made sure not to miss a word of Dimitri's description of this apocalyptic past. 'This civilisation must have been a fragile one indeed to have collapsed in such a short time...'

'Well, not exactly. This civilisation was actually born in the late Middle Ages, in the Thirteenth century. As Twentieth century political scientist Carl Schmitt noted, it bloomed in the Sixteenth century, in the age of "great discoveries", when Europeans set off to conquer other continents. Its peak can roughly be situated between 1860 and 1980. Still, already in 1921 – about one century before its end – a German philosopher, Oswald Spengler,[13] had seen the first signs of the future collapse. This civilisation lasted seven centuries – a little less than the Roman Empire. As is the case with all civilisations destined to collapse, its end was very close to its peak... for the "viruses of decline", after being at work invisibly for a period, tend to suddenly become deadly when a civilisation has reached its highest peak.'

'You seem to be obsessed by "catastrophe theories"!'

'I'm not obsessed by these theories. These are laws that explain the course of history, as well as many other phenomena. The worm may already be in the fruit, but the fruit looks attractive. The old oak might be at the height of its vigour, but it is rotting inside and will be uprooted by the first storm.'

Dimitri suddenly added, 'Fasten your seatbelt, Nafissa. We are about to slow down – we've reached Novosibirsk.'

Dimitri continued his improvised history course. 'Between 2018 and 2020, the rest of the world also plunged into chaos.'

'How?'

'The global financial system and stock markets continued to fall, and environmental and climactic disasters did not abate. In two years, the depletion of fishing resources, the impoverishment of the soil and desertification caused a terrifying series of famines. It is estimated that by 2020 two billion people had died...'

'Who resisted?'

'Paradoxically, Russia kept going. This is very important for the rest of my narrative. Russia had been the "sick man of Europe" in the late Twentieth century, following the collapse of Communism. But the new and largely militaristic regime enabled the country to resist. Your country, India, also resisted, as did China and Japan. These areas preserved their unity, as they were ancient civilisations that hadn't forgotten their archaic self-defence mechanism. Despite the huge crises, they preserved their political homogeneity and technological economy, which slowed down but still functioned. By contrast, multi-ethnic

---

13 Oswald Spengler (1880-1936) was a German philosopher and is considered part of the Conservative Revolution of the Weimar era. His most important work was *The Decline of the West*, in which he theorised that all civilisations go through an inevitable cycle of ages of rise and decline in power. Spengler saw the West as entering its period of decline at the time he was writing.

societies in which traditions had been destroyed or marginalised to make room for an economic cult imploded, for they no longer had any social or political thread holding them together. This is what happened in Western Europe and North America. But it is interesting to note that this global hurricane and pandemic spread from France: the country of the Revolution and philosophical birthplace of modernity was the first one to commit suicide. Poison always affects the head first...'

After some silence, Nafissa asked, 'When, in 2017, the Muslim army entered France, why didn't neighbouring countries try to defend it? Weren't they all part of this "European Union"?'

'They didn't intervene because of cowardice, although this is not the only reason. Since 2014, the European Union had been little more than a fiction. The various European armies practically no longer existed, nor were they motorised. In these conditions, how could they ever have faced a North African and Muslim army equipped with fuel, armed vehicles and resolute leaders?'

Stop in a Tunnel

Nafissa didn't answer Dimitri. Suddenly, there was a violent tremor. An artificial female voice confirmed the information flashing across the screen in front of them. 'The train has come to a halt because of a minor accident. We shall keep you updated.' The train shook as it braked abruptly.

'This is quite normal at this speed. Accidents often occur when the train is braking. I just hope I won't miss my connection in Komsomolsk with the airship for Bering.' (Dimitri's voice betrayed a certain anxiety.)

The lights dimmed inside the car because of the loss of electric power. The computer screen on the seat in front of them switched off. Things were getting rather unsettling...

Nafissa smiled. 'Don't worry, Mr. Councillor. The gods of ancient India will protect us.'

She started laughing and shook her black hair. 'What a sorceress,' Dimitri thought. An almost perfect silence reigned inside the car, which was enveloped in the half-darkness. Passengers' whispers could be heard, faintly. An ill-boding smell of burning filled the air...

A DAY IN THE LIFE OF DIMITRI LEONIDOVICH OBLOMOV 219

They were stuck like prisoners in a tunnel in the heart of Siberia, under the taiga... Dimitri could picture the forest of pines and birches above them: alternating black and white trunks and rippling streams in the woods. He closed his eyes. Five metres above their heads, he imagined a carefree *moujik*[14] dressed in fur and leather with his dog and an iron-tipped, gnarled old stick gathering dry twigs to light the evening fire in his *isba*.[15]

With a very Zen attitude, Nafissa continued with her questions. 'So how did liberation come about in the end? Why don't there seem to be any more Muslims in Europe now? Please answer me and relax. Take some deep breaths...'

Dimitri did so, and Nafissa rested her sweet, warm hand on his. Again she urged him, 'Relax. My gods will protect us. Now please answer my question.'

Dimitri spoke in a low voice, 'In 2025, the "baronies", or areas of European resistance, which were living as if under siege, chose to ask for help from the nationalist, populist Russian Federation. What led to this decision was the Muslim conquest of the free state of Lorraine, which included the city of Metz and its surroundings. The acts of reprisal the Islamic army carried out there were atrocious: the city cathedral was burnt down and the Russian ambassador was slain along with all his family in reprisal for the anti-Islamic policy that Russia and the Orthodox Slavs had long adopted.'

'So Russia launched a sort of Crusade, but this time to the West?'

'Yes. I see you actually know history quite well, Nafissa. So on 6 June 2025, which is now celebrated as the day of the 'Proclamation of the Reconquista', General Alexander Ivanovich Dukachevsky, the Lord of Russia, accepted the plea of the besieged cities of Europe. In December 2026, an army of over a million men backed by tanks and jets crossed central Europe into the "Western Europe occupation zone", which included France, Spain, Italy, Belgium and Holland, as well as portions of Germany and Scandinavia. A second army of 300,000 men from Ukraine, Poland, the Baltic, Finland, Serbia and Greece, as well as troops from the states under Russian "protection", landed in Brest. Here they joined the Breton army – 80,000 men strong – and marched to the East in such a way as to squeeze the Islamic forces like a vice. The Russians

---

14 Russian: 'peasant'.
15 A traditional style of house in Russia.

provided fuel and ammunition. The decisive battle took place in the Brie area, east of Paris, near the ruins of a huge Twentieth century American theme park.[16] Most of the Muslim army was destroyed and the survivors were made prisoners. A second battle took place in the Maurienne valley in the Alps. The victory of the liberation troops can be explained on the basis of two factors: on the one hand, the Muslim troops were badly organised and suffered from inner divisions resulting from quarrels among their leaders; on the other hand, the Islamic republics, which had been hit by the global crisis, could not afford to provide them with ammunition and fuel. It was no longer an organised military force: it was more like a horde. The victorious army entered Paris and was cheered by the meagre population still living there (the city had been nearly abandoned). What followed was the "Reconquista of 2025-28", which was unfortunately an extremely violent one.'

A humming noise was heard. Suddenly, the lights came back on and the screens on the back of the seats switched on again. An artificial voice announced, 'The accident has been repaired. The damage was caused by an electro-magnet that had caught fire. The light smoke will be cleared by the air conditioning. We will be running eight minutes late. Trans Kontinent Ultrarapid apologises for the inconvenience caused. The connecting airships will be waiting for all passengers. Thank you.'

'You see, Mr. Councillor. I told you everything was going to be OK.'

Nafissa withdrew her hand. The train took off again, sliding along at a reduced speed (450 kilometres per hour) down the tunnel before coming to a halt in Novosibirsk station.

Novosibirsk-Irkutsk

The bullet-train stopped for three minutes. It then departed again in the direction of Lake Baikal. The screen said, '13,000 kilometres per hour. We are reducing our delay to two minutes.'

Nafissa went on, 'Why did the United States not intervene as they seem to have done with other invasions in the past, such as when they freed Europe from those merciless German dictators?'

The Indian girl had a naive attitude. The Russian councillor smiled and answered in a professorial tone, 'The reason is very simple: the United States no longer had any means at its disposal. And besides, it

---

16 This would be Euro Disney, or Disneyland Paris.

had no wish to free Europe from the Islamic yoke. It had other things to worry about! Following the huge global economic crisis I mentioned, the United States imploded. It had been the leading world economic power, but its unity was only based on widespread economic wealth and financial investments. From 2020, people in the United States started fleeing from the cities, as was happening in Europe, but for other reasons: the increasingly impotent federal state disintegrated, the economy came to a halt and famines and epidemics broke out, as well as ethnic conflicts – such as the terrible clash that took place between Hispanics, Blacks and Asians in October 2020 in Los Angeles. The same scenario occurred as in Europe: 35% of the population disappeared, as states proclaimed their independence and withdrew into themselves. Blacks regrouped in the South and Whites fled the areas in which they were a minority. A new ethnic map was drawn in this vast area. Only two regions managed to keep their industries and economies running, if only at 20% of their former capacity: the American Republic of the Pacific, situated on the coast between San Francisco and Vancouver, and which became a sort of Sino-Japanese protectorate (and remains such to this day), and the Old American State (OAS), which stretched from Michigan to New England and included southeastern Canada, with Chicago as it capital...'

'And what about New York, this legendary ancient city?'

'All that remains of it today are huge ruins that can be visited...'

'I know,' the Indian girl replied. 'My father, like all high-ranking officers in our Empire, received an advertisement on his video-programme that said: "Don't miss the fantastic view of New York's ruins." It was an offer from the Indian Tourist System to get a view of the remains of the city from an airship.'

'I see... Right from the onset of the economic crisis, New York turned into a living hell. With the rise of the sea level, at each great tide it was devastated by floods. Riots, fires and famine did the rest. New York lost all its population in a very short time. As you know, there is no such thing as a "short-acting catastrophe". Catastrophe theory speaks of a "final acceleration". This is the famous law of 80-20: 20% of a system will collapse in 80 units of time and the remaining 80% will collapse in 20. New York, a symbol of modernity worldwide, could not survive its brutal end. I should add that Los Angeles, as you know, met the same fate as New York...'

'Yes, I'm aware of this. But apparently the ruins of Los Angeles are far less striking when seen from an airship.'

'Well, that's because they were mostly knocked down by a huge earthquake in 2043. There were hardly any victims, though: the area had already been abandoned.'

Dimitri's computer made a beeping sound. He typed '18' on his keyboard, to enable the flow of information. Suddenly, Vega showed up on the screen. She had changed her dress and was now wearing an ancient Greek *peplum*.[17] In the background, a Greek pipe was playing a languid song from the 1970s, '*Millisé mou hos agape mou*'[18] – an incessant, ternary motif from ancient Thessaly.

Nafissa broke out in a laugh. 'You certainly did a good job in designing your virtual secretary! It really matches your fantasies, Mr. Councillor! I hope your wife doesn't know about her...'

Dimitri mumbled, 'Of course not. This super-powerful quantic computer is for high-ranking officials only. Surely someone my age has the right to a little fun?...What's going on, Vega?'

'Master, the Supreme Inter-State Court of St. Petersburg wishes to inform you that the Kingdom of Albania is asking for a two-year delay in the repayment of the debt it contracted with the Federal Bank and the Republic of Kamchatka in 2070. They are anxiously waiting your verdict.'

Dimitri typed on his laptop, 'Grant them a 16-month delay – no more than that. If the Albanians don't accept these terms, the Federation will consider revoking its funding for the wide canal between Tirana and Sofia. I'm fed up with these good-for-nothings.'

The computer remained silent for a moment. Then there was a hiss. Vega's image remained motionless, before coming to life again.

'Should I write "good-for-nothings" in my answer to the Court, Master?'

'No. Delete the last sentence and rewrite the whole thing in administrative jargon.'

Dimitri typed '81' and the image of his virtual secretary disappeared. The Indian girl had witnessed the whole scene.

'You make decisions fast...'

---

17 A *peplum* was a body-length women's garment.
18 'Tell Me that You Love Me'.

Dimitri felt flattered and answered, shrugging his shoulders. 'I have to. The Federation includes 125 autonomous states, each of which has its own egoistical demands. The rule of general consensus can no longer be applied, as it was in the Twentieth century. Decisions must be made, in the name of the Imperial Government and the common interest.'

'What if a state does not agree with your decisions?'

'It can hold a referendum and leave the Federation. This is what happened with the tiny state of Corsica, with Euzkadi or 'Basque Country', with Sicily, Estonia and others, too. Some of these have now made their way back into the Federation, while others are begging us to accept them. This is quite natural, as they no longer benefit from our federal solidarity and military protection.'

'We've faced exactly the same difficulties in the Indian Empire. Nepal first left the Union but then joined it again out of fear of China...'

'With the states of Brittany, Bavaria, Flanders, the Ile-de-France and Sweden, we've had the opposite problem: they are highly dynamic and are trying to get their hands on everything. They make their presence felt in all the ministries and commissions. The worst of all are the Bretons. They're everywhere. They would make you believe it's they who are governing the Empire. Not that this is far from the truth... The current President of the Imperial Government, our head of state, is a Breton.'

Nafissa stared at Dimitri in amazement. He added, 'Well, despite superficial disagreements, there is an understanding between us, as we've all realised that we're part of the same folk – although there are over 20,000 kilometres between us. Disputes about selfish economic interests are part of life. What ultimately matters is agreeing about the important issues.'

'And what are the "important issues", then?' Nafissa asked in a mischievous voice.

'Identifying our common enemies – and our common friends.'

'Oh, I see. I pretty much agree with you.'

The girl then changed subject. 'You were saying that the 'Reconquista' which took place between 2025 and 2028 was a very brutal one... Would you tell me more about it?'

Irkutsk-Komsomolsk Terminal

Dimitri could detect a taste for tragic stories in the Indian girl's eyes. She fastened her seatbelt. The deceleration was very sudden. The screen said '3.2 G'. The train stopped in *Irkutsk* for less than two minutes. A man with long hair and a vermillion uniform took a seat near them, at the other end of their row. Along with his travel bag, he was carrying a pinewood easel. Dimitri realised this man was a Lieutenant Colonel of the 2nd Imperial Artists' Battalion. His collar badges – silver ones on a mauve background – were adorned with a crossed paintbrush and hammer. The train set off again.

Dimitri gave a belated answer to Nafissa's question. 'Yes, it was very brutal. After the Great Catastrophe, as is always the case in history, people's system of values had crumbled. It was General Dukachevsky who took things into his hands. The remnants of the Muslim army and the ethnic gangs were captured and assembled in the south of what had once been France, and then forcibly shipped to North Africa, which had no military means to oppose this operation. But something even more serious happened: because of the traumas they experienced and the radical changes in their outlook, all the descendants of the great waves of extra-European immigration that had hit western Europe, in particular since the 1960s, were unfortunately... well, *deported*. We are talking here of several tens of millions of people. You can well imagine how this operation carried out by the "European Liberation Army" was no gentle business... This is what historians call the "Reconquista".'

Beautiful Nafissa looked at Dimitri in surprise. 'Why did you say "unfortunately", Mr. Councillor?'

'I find all these events rather shocking from the point of view of my own conscience and my old Christian upbringing – but so it was...'

'As a Hindu, I'm not at all shocked. Well, please continue: what happened then? Did massacres take place? Is this what you deplore?'

'No, there were no massacres. These rootless people without a homeland were transferred en masse from Europe to the island of Madagascar by boat. There were 23 million of them. Many were legally 'French', 'Belgian', 'Dutch' and 'British'. But this meant nothing now. The nationality rights of the old world had completely disappeared... Archaic criteria had come to prevail.'

Nafissa gave a wide-eyed look of amazement. 'In India they never told us anything about this....'

'The Government financed the whole operation. The Federation is currently paying the Kingdom of Madagascar 10 billion Eurosesterces. Integration has worked very well down there. '

The Indian girl posed a new question. 'How did Twentieth century science and technology survive the "Great Catastrophe"? How did humanity manage not to plunge into primitivism again?'

'As was the case after the end of the Roman Empire, "pockets of survival" had endured, as if by a neo-medieval reflex. And besides, India, China and Japan all resisted far better than the West. The collapse was contained. Most of the technologies that had been acquired were not lost. Technological expertise was "frozen", not abandoned. Innovation came to a halt, but minorities spared from the general chaos somehow ensured the transmission of knowledge in just about every corner of the world. This made the Second Renaissance possible, which took place around 2030.'

'Tell me about it...' Nafissa changed the audiocassette in her recorder.

'Between 2030 and 2038, the various "baronies" established mutual contacts, as communication had become possible again and peace had been brought to their lands. A spontaneous regrouping into "autonomous states" then took place in Europe and the continent restored its old capital, Brussels, yet this time on the basis of principles utterly different from those of the former European Union. Nation-states, such as France or Germany, were never re-established, as they no longer inspired any trust in the people. This new form of organisation, which was at first called the Community of European States, included the ancient regions of Western Europe – Bavaria, Wallonia, Padania, etc. – which were largely autonomous.'

'So how did you go about creating this huge "Eurosiberian Federation" you also refer to as "The Empire"?'

'By 2038, the economic system had been restored, although it only produced 10% of the goods and revenues it had been churning out before 2014 – and no one wished to produce more. Everywhere the countryside was repopulated. A minority living in small cities took up an ultra-scientific way of life and soon improved upon Twentieth century discoveries. Still, great international problems soon surfaced again, with the risk of nuclear and bacteriological warfare. The Islamic Republics, your own country (the Empire of India), China and Japan, among other states, were involved. Russia and its central European satellites then invited the Community of European States to simply merge with them, in order to ensure the unity and defence of their 'kin peoples'. This took place with the Pact of Prague, signed in December2038, which solemnly signalled the establishment of the Eurosiberian Federation. This union immediately resolved these international tensions... After two years of difficult negotiations, in 2040 the institutions of what we now refer to as our 'Great Homeland' were defined. In this same year, work began on the first line of the planetrain we are riding on...'

The train suddenly decelerated and started shaking as it slowed down. On the screens a red light started flashing.

'Fasten your seatbelt, Nafissa!'

'What's going on? Is it an accident?'

The girl didn't look at all afraid, although she acted as if she was. Dimitri lightly touched her hand on the arm-rest. Nafissa withdrew it immediately.

'No, don't be afraid. Between Magocha and Skovorodino the planetrain track is no longer underground. So the train must slow down, and it does so suddenly. We're no longer in the vacuum of a tunnel, but in the open air.'

An icon popped up on the upper part of their screens for a few seconds that read: 'Train slowing down. Above-ground track. Speed reduced to 420 kilometres per hour.'

Dimitri cleared his throat and explained, 'In this region the nature of the soil prevents the digging of tunnels. The planetrain is losing speed because of air resistance. Look...'

With a buzz, a panel slid up electronically, revealing the window. Daylight flooded into the cabin and the electric lights went off. The young Indian girl leaned over towards Dimitri to get a better view of the landscape beyond the small Plexiglas opening.

Magnetically suspended on its large, elevated monorail, the train rolled on across a landscape of forests, misty mountains and boundless horizons – the landscape of eastern Siberia, straight out of a Tarkovsky[19] movie...

'Look!'

---

19 Andrei Tarkovsky (1932-1986) was a great Russian filmmaker of the Soviet period. His films frequently depict the Russian countryside in a mysteriously pastoral manner.

The forest had disappeared. The train was now crossing a huge city made of wooden houses, huts and *izbas*. A brick Orthodox church could be seen, surmounted by a golden dome, followed by a cattle fair and a laundry-house crowded with women. Despite the train's speed, passengers could make out markets full of people, horse-drawn vehicles, fields ploughed by oxen, farms, the banks of a great river dotted with watermills...

This sight lasted several minutes. In the distance, huge ruins could be glimpsed covered in vegetation – the remains of industries and carcasses of buildings: the old mining town of Magocha, a vestige of the Twentieth century. Beyond it was virgin nature with its endless forests of pines and birches.

Dimitri went on, 'This is one of the largest neo-traditional communities in our Federation. There is one airship a week connecting it to either Ulan-Ude or Irkutsk. My wife, Olivia, visited the place last month to buy some smoked yak meat and *vodschkaia,* a wonderful liquor made from birch bark which cannot be found anywhere else. This community has at least 50,000 inhabitants. They have more or less the same standard and way of life as the people of Thirteenth century Europe. They are very happy as they are...'

'Is it true,' the Indian girl asked Dimitri, 'that before the Great Catastrophe they had tried to make all of humanity adopt a form of economy based on technology?'

'Yes, that was the great utopia of the Nineteenth and Twentieth centuries. It originated in Europe and America, but it was not at all viable. It contributed to the collapse of the old civilisation and to the migration of people from South to North. Today only 19% of the inhabitants of the Eurosiberian Federation partake in the techno-scientific economy and way of life. I believe it's even fewer people in India...'

'In my country only the caste of the Abishamis,[20] to which I belong, lives that way. I think we make up about 5% of the population, which still means tens of millions of people. And in any case, according to what my father the Maharaja of Gopal says, society is far more just and balanced today than it was in the old world. India has rediscovered its traditions.'

Dimitri smiled. 'Clearly, it's no longer a "democracy"...'

Nafissa raised her eyebrows.

'What's "democracy"? I've already heard this word...'

---

20 In the Hindu caste system there is no element by this name. It may be that Abishami is a family name that is associated with a particular caste.

This was the sort of question that disturbed Dimitri, so he tried to give an evasive answer...

'Initially democracy wasn't a bad idea. In ancient Greece it meant power to the *demes,* or townships. But then it spread to all peoples, including in very populous countries, and this cultural grafting proved disastrous. Democracy only fits the European mind-set. It cannot be exported: each folk has a specific mode of government to which it has adapted itself. When badly applied, democracy can lead to injustice and disaster or become a front for tyranny.'

'I really cannot understand how Westerners could have believed that the whole of humanity was meant to live according to the same regime. What lack of common sense and what pride!... In India we're not "democratic", but ours is not an unjust or tyrannical system and all works well as it is...'

After a moment of silence, Nafissa added, 'And what about the Federation, have you re-established "democracy" within it?'

Dimitri gave the girl an ironic smile.

'Let's put it this way: we no longer have the kind of democratic system that was in force before the Great Catastrophe. We are now applying the notion of "organic democracy" inspired by Plato, an ancient Greek philosopher. A fixed and uniform model of democracy would be completely unfeasible for a geographic entity such as ours, in which vast differences exist between the way of life in rural communities and that of the minority of people, like the two of us, who have resumed the techno-scientific lifestyle. Besides, each of our autonomous region-states is free – in all those areas which are not within the province of the Imperial Government – to organise its institutions as it wishes. All these states have to do is appoint – by whatever means they please – a fixed number of representatives for the Federal Senate of the Empire, in proportion to its population. But rest assured: no state has the right to oppress its population, lest it be expelled from the Federation. The state under the rule of law is our norm.'

Nafissa gazed at him intently with a half-smile.

'I understand. You really are very tolerant! My father would have a good laugh! But then again, every folk has its rules... Please continue with your explanation.'

Dimitri didn't react.

'In the Federation we have tried to combine two principles: on the one hand, absolute authority and quick decision-making on the part of the leading political body – the Government elected by the Imperial Senate; on the other, great freedom of organisation for individual region-states. Some of these – about 30% – have remained or turned into hereditary monarchies ruled by kings, dukes or other rather folkloristic sovereigns. As you see, we try to be both tolerant and efficient.'

* * *

The stewardess interrupted their conversation to serve them some cubes of raw fish from Lake Baikal, mixed with hot vegetable soup – a typical dish from the area they were crossing. Nafissa was ravenous and devoured her food.

'Your cuisine is good, Mr. Councillor. It's almost as tasty as that of my own country.'

'I organise culinary competitions between the various autonomous states of the Federation on a regular basis.'

'And who wins?'

'It's annoying: it's always the states from old France...'

'Open the competition up and then the Indian Empire could take part...'

After a moment of silence, Nafissa said, 'Look!'

She was again leaning against him, her face against the window. Her long black hair brushed against Dimitri's uniform. He focused on the view outside.

The train was moving between a rock face and a clearing. Dozens of animals with grey fur were running in the undergrowth. They only caught sight of them for a few seconds.

'It's a pack of wolves. They're multiplying everywhere. In the Twentieth century wild animals had disappeared, but now they've made a big comeback. Clearly, this is causing quite a few problems...'

'It's the same with tigers in India. From time to time they devour a villager. But they're so beautiful! Look, I've got a bag made of tiger fur...'

'I've seen it. I've recently had to settle a controversy between the Duchy of Provence, the State of Padania and the Federal Ministry of Agriculture. They were complaining about the proliferation of wolves, which destroy flocks, and asking us to send them 5,000 trained dogs to protect them. But the cost of this was too high and the negotiations dragged on.'

'What was their outcome?'

'The two states have 25,000 shepherds with huge flocks and I came up with a brilliant idea to solve the problem.'

'I'm not surprised... Tell me.'

'At my request, the AHG (Animal-Human-Genetic) labs, a branch of the huge Typhoone company, developed 1,500 "biotronically modified animals", two for each threatened herd. These were far cheaper than 5,000 trained shepherd dogs.'

'What are these biotronic animals?'

'They're biological animal-robots: genetic hybrids of various species, including man, of which wolves are naturally afraid. They are crammed with electronic chips that multiply all their faculties tenfold and enable them to do without sleep. So at night they remain awake to guard the flocks. Clearly, no wolf dares approach them now...'

'And what do these things look like?'

'Pretty much like the gods of Hindu mythology!'

Nafissa frowned. Dimitri continued, 'Oh, I'm sorry! Well, they walk on two legs, have huge limbs and a head that is halfway between a monkey's a shark's... They look a bit like a dinosaur from the Jurassic era, the Velociraptor. They're guard animals equipped with exceptional bodies and there's no need to train them because they're already programmed beforehand. Their cost has considerably dropped, as the AHG has decided to sell a modified version of them to the police forces of the Federation's states and to the Federal Army. Clever, huh?'

'Indeed... This lab should canvass the Indian market. But tell me, Mr. Councillor,' Nafissa remarked in a flattering voice, 'you must have some great responsibilities...'

'Well, my job is both very simple and very complicated: I must settle disputes among autonomous states and make everyone respect the laws of the Federation. I command 2,000 federal officials – if you really must know,' Dimitri added, stroking his epaulette.

'I'm happy for you, Mr. Councillor. My father, the Maharaja of Gopal, has about a hundred times as many people under his control!'

Nafissa broke out laughing, as Dimitri scowled in his corner.

The train continued travelling on its elevated rail and sped across a deep forest.

'What's that?!'

The girl again pointed to something beyond the window. She had glimpsed a bizarre object shining in the sun on the tops of the pines, behind a slope.

'It's a "Barge" of the FAF, the Federal Armed Forces.'

The object was large, oblong and parallelepiped, slightly curved at its ends, measuring some twenty metres in length. It looked vaguely like a flat-keeled river barge. The object was fluctuating and spinning on its axis as it surveyed the forest. It was khaki green in colour and both its sides and bottom appeared to be covered in tubes.

'What's that machine for?'

'That "machine", my girl, is one of the most promising new inventions by Euromotor, a rival company of Typhoone. The principle behind it is this: the Barge is made of a new super-light material, keflon, which weighs less than cotton but is as resistant as steel. It floats in the air because a vacuum is created at its centre. It is piloted from the ground and moves thanks to neutron-based mini-reactors. It is equipped with radar, 3D cameras and a miniaturised and highly sophisticated electronic detection system.'

'So it's like a flying radar?'

'Exactly. But it's an extremely accurate one. It is used to discretely identify all possible threats, from local to wide-scale ones. It is far more efficient than the old radar-planes. It can fly between 10 and 15,000 metres without making a sound and is difficult to spot. The Typhoone company recently announced that it is perfecting a new generation of rival Barges which perform even better as they are based on an anti-gravitational system...'

'And what's that "Barge" doing there?'

'It's probably a military operation or some kind of experiment,' Dimitri answered evasively. 'In eastern Siberia such things are quite common because China's so close.'

The girl's voice took on a more perfidious tone. 'So, Mr. Councillor, is the vast Eurosiberian Federation planning to go to war? And against whom?'

'Don't believe it, Nafissa! Twenty-first century history has made us peace-loving but not pacifist. We simply wish to make sure that no one will ever be able to attack, invade or defeat us. Our aim was to create a federal army that no one would dare face. The only goals of the military policy of the Government are to protect our "common home" and prevent human folly from destroying the planet – deterrence based on the potential threat of our power. But don't worry: we have no intention of attacking anyone, and certainly not your marvellous civilisation... In this respect, we're following the ideas of de Gaulle and Gorbachev.'

'And who are they?'

'Oh, they're European heads of state from the mid-Twentieth century. Hardly anyone listened to them in their day...'

Komsomolsk

The train entered another tunnel. The lights came on, the windows were covered and passengers were pushed against their seats as the train gained speed. The screen read: 'A speed of 12,000 kilometres per hour will be reached within seven minutes. Acceleration level G3. If feeling unwell, contact your stewardess.'

A few minutes later, the train had reached its terminal, the underground station in Komsomolsk. Dimitri took bitter leave of the Indian girl. The two exchanged their electronic coordinates on the platform.

'Call Vega to get in touch with me,' Dimitri said with a note of sadness. 'My invitation to Brussels is always open.'

'And you *and your wife* are welcome to visit my father's palace in Srinagar.'

'Where are you headed, Nafissa?'

'I will continue my study trip. I'm going to China. The Governor of Manchuria is a friend of my father's. I've booked a sleeper car on a classic train for Changchun.'

'It will be a very long ride – 1,200 kilometres, more or less...'

'Oh, it will be very comfortable. And I've got all the time I need. Besides, I've brought something to read: *Trifles*, an English novel from the Twentieth century. It tells a horrible story: the voyage of the *Titanic*; this time, though, the ship doesn't hit an iceberg but safely reaches New York.[21] It describes frightful things. I love it.'

Dimitri's gaze followed the slender figure of Nafissa, her hips swaying, as she disappeared into the crowd. With her dark skin, she was almost a living version of his virtual secretary, Vega. Would he ever see her again?

---

21 There is no actual book by this title or description, although the idea of the *Titanic* successfully completing its maiden voyage has been explored in a number of real alternate history stories.

* * *

Dimitri took an escalator and reached the surface. He had left Brest just over three hours before, early in the morning. Now, because of the time difference, it was already night. The cold hit him – it was only 10 degrees Celsius, despite it being already June. Eastern Siberia had hardly benefited from global warming. The skies twinkled in the black sky. Outside the station building, the fires of the chestnut and grilled fish merchants glowed.

There were no electro-taxis here... Dimitri headed for the carriage station. There was a queue of about twenty people waiting. He entered the stationmaster's cabin and showed his high federal official's card. His uniform would have been enough...

'Please follow me, Mr. Councillor...'

A few minutes later, Dimitri was sitting in a carriage pulled by a small black horse with a lively trot. On the seat was a small heater containing some burning coal.

'Where shall we go, sir?' the coachman asked in Cumikan dialect.

'To the airship harbour. Hurry!'

The coachman cracked his whip and the horse quickened its trot.

Once jets had been abandoned as a means of civilian transport and replaced by airships, airports no longer required long landing strips, nor were they a source of pollution and noise. So they could now be built rather close to city centres. It is often updated versions of old technology that prove the most efficient, as was the case with the new wind-based systems of propulsion used for ships. Airships were not as fast as the old jets, but helped gain time in the final lap of one's journey.

Dimitri had booked a first-class ticket on the airship that directly connected Komsomolsk to Dorbisk, his home on the Bering Strait, 2,300 kilometres away to the northeast. The airship also made a stop in Petropavlovsk, in Kamchatka.

After a ten minute, bumpy ride, the carriage stopped outside the station run by Siberik Sideral Flot, the public company owned by the United Republic of the Siberian Far East. There was no way Dimitri could have paid with his credit card here, so he gave the coachmen a two-Eurosesterce silver coin.

It was an amazing sight: a dozen mooring masts, measuring a hundred metres in height, stood there, lit under the starry sky. At the summit of each mast a huge, black cylinder lightly danced in the wind, its position lights on. These were the airships.

A blonde stewardess with violet eyes led them to the base of the mast for flight 788. Dimitri boarded the aircraft via a lift and stored his travel bag in the hold, keeping only his laptop with him. He took his reserved seat, by a window. It was even more comfortable than the planetrain. There was a screen in the back of the seat in front of him and a light meal on the tray, including a glass of *vodschkaia,* the liquor made from birch wood. As they waited for departure and the propellers of the airship started to turn, SSF advertising flashed on the screen. An electronic voice accompanied the text:

'Thank you for choosing our air travel company, the safest in the world. We ensure connections across Siberia, with departures from 80 cities, serving 35 rural communities. This airship is an Albatross 350 built by Typhoone. It is driven by six propellers with different blades fed by a fast-neutron nuclear reactor. Its level of atmospheric pollution is zero. We are supported in the air by two systems: a helium compartment and a vacuum one. The aircraft holds up to 200 passengers, including 50 in first class. A bar and prayer chapel are available for passengers on the first floor. We will travel at an average altitude of 3,500 metres and with a speed of 490 kilometres per hour. With favourable winds, we can reach a top speed of 580 kilometres per hour. We will land in Dorbisk, our final destination, in just over six hours. The Commander, Markst, and his crew would like to wish you a pleasant journey. We are now ready for takeoff.'

On the Airship

There was a humming noise from the reactors and then the huge aircraft, unmoored from its mast, set off at high speed, flying over the city and its lights. The airship then gradually veered left. Dimitri looked out of the window into the darkness. They were already crossing the Sea of Okhotsk. The cabin was flooded by a bluish light.

It was now time for Dimitri to get to work. He switched his computer on and connected it to the screen in front of him. Vega immediately popped up. This time, Dimitri's virtual secretary was wearing

A DAY IN THE LIFE OF DIMITRI LEONIDOVICH OBLOMOV 235

a tight-fitting, long muslin[22] dress, in an early Twentieth century style.

'I am listening, Master. I have just received a message from Commodore-Count Ron Schneider. He is waiting for your report and is growing impatient. He is complaining that you have switched your mobile phone off... He cannot get directly in touch with you.'

Dimitri had switched his phone off so as not to be disturbed when speaking with Nafissa on the planetrain. A small lapse in professional etiquette...

'It's no big deal. Make a note of my report, which includes a recording of the most important moments of the Brest conference, and send it immediately to Schneider in St. Petersburg, via Euronet.'

This means of communication, which had already been known in the Twentieth century, had entered development in 2010, before the Great Catastrophe brought things to a halt. Only around 2050 was this technology resumed – and improved – thanks to the superior power of quantic and bionic computers ('ADN chips'), although clearly it was reserved for the ruling elite alone.

Dimitri started dictating his report over the microphone. His words were immediately being transcribed (and translated) in the form of a trilingual text (in Russian, French and German) by the computer and would then be sent via satellite to Schneider's fax machine. Dimitri was then going to insert the microdiscs containing the recordings of the debates into the disc-reader on his laptop. These, too, would then be immediately transcribed as texts and attached to the report, which would reach Schneider in the headquarters of the Supreme Inter-State Court within less than a minute.

Dimitri grabbed the microphone and started speaking in a low voice, so that the other passengers could not hear him.

'Are you ready, my beautiful Vega?'

'I'm ready, my wise Master...'

Dimitri's virtual secretary shuffled on the screen, with a killer pout on her face. He had programmed her very well indeed... And to think that this dream girl didn't really exist!

'What follows is the introduction to the report.' Dimitri was speaking slowly and the transcribed sentences flashed on the screen in

---

22 Muslin is a type of cotton fabric first introduced into Europe from the Middle East.

Russian. Typing on his keyboard, he would change a formula or expression here and there.

'From Dimitri Leonidovich Oblomov, Plenipotentiary Councillor, to his Excellency Commodore-Count Ron Schneider, Provost of the Supreme Inter-State Court of the Eurosiberian Imperial Federation.

Object: Settlement of a dispute among the following autonomous region-states: the Republic of Ireland, the Kingdom of Scotland, the Duchy of Wales, the Duchy of Cornwall, the Popular Republic of Brittany, the Duchy of Normandy, the Free State of Vendée-Poitou-Charente, the Duchy of Aquitaine, the Socialist Republic of Euzkadi, the Republic of Galicia and the Federated States of Portugal and Lusitania – all of which are all members of the Association of Economic Interests known as "Atlantic Arch" and represent the respondent party. To these states is opposed the accusing party, comprised of: the Kingdom of Île-de-France, the Socialist Republic of Ukraine, the Kingdom of Bavaria, the Unified State of Padania, the Kingdom of England, the Czech Republic and the National-Popular Republic of Serbia.

Nature of the dispute: The aforementioned accusing autonomous region-states (ARS), the plaintiffs in the case, are charging the aforementioned respondent ARS of the Atlantic Arch of having acquired a de facto monopoly over the fish market, fishing reserves and farming of shellfish and seafood. As high yields allow these states to keep fish prices low, they are damaging the agriculture of the accusing ARS by providing unfair competition for their exports in the Federation, causing them great economic losses. According to experts, this complaint is well-founded. The aforementioned accusing ARS are asking for the states of the Atlantic Arch to provide financial compensation by subsidising their produce. The latter states have refused. My mission was to find a solution by enabling these states to reach some sort of agreement.

Location of the meeting: Federal Navy Ministry, Brest, Popular Republic of Brittany, 20 June 2073.

Participants in the meeting: 1. The presidents of the parliaments of the aforementioned autonomous region-states; 2. two experts from the Federal Financial Delegation; 3. myself, your servant. The meeting was chaired by Father Wencslas, President of the Republic of Lithuania, a state not affected by the economic conflict.'

'I'm first sending your Excellency the recordings of the most interesting moments of the negotiation.'

Dimitri inserted the disc with the recording into his computer.

'First, there's an exchange between Mrs. Gwen Ar Pen, President of the Parliament of Brittany, and myself...'

'There's no way we will ever fund the produce of those agricultural states! All they have to do is be productive, like us, and come up with innovations so they can export their rabbits and sheep at a cheaper price. I also wish to point out, Mr. Plenipotentiary Councillor, that the Breton state is an agricultural power, too, and that we manage to export our pigs, fruits and vegetables at competitive prices! We scrupulously respect the federal law imposing organic farming and banning industrial produce and GMOs.[23] If a Breton pig is 50% cheaper than a Czech one that's because we're better organised. Our neighbour to the south, the Free State of Vendée-Poitou-Charentes, which is also a maritime and agricultural state like us, does the same with its butter and spirits. The states of the Atlantic Arch also respect the federal laws banning the use of trawl-nets for the preservation of fishing resources. As a solution to the problem, I suggest the Federal Financial Delegation from Frankfurt fund the agricultural produce of the plaintiff states. I'm certain the latter will gladly accept such a solution.'

'Madam, this is quite impossible. In this case, we should also be funding the agricultural produce of all 125 autonomous states of the Federation, so as to avoid making anyone envious! And this is not a realistic financial prospect. Besides, it also goes against the principle of financial responsibility for the Federation's states. Let us not forget that the federal budget is already completely funding the rearing of workhorses and the spatial programme of low-orbiting nuclear plants, as well as the Hipparcus Crater mining base on the Moon, a space programme which was initiated by your own state – I would like to remind you – in partnership with the Republic of the Flanders, Bavaria and Moscow, amongst others. You are autonomous and cannot expect everything from the Federation. It seems to me that the Breton state is often taking too many liberties with respect to federal agreements...'

'Such as, Mr. Councillor?'

'For instance, why is it that the Breton language is so badly taught in Brittany? You are contravening to the norms of decree R.567 of the

---

23 Genetically modified organism.

Language Committee. Each nation of the Federation must teach its subjects its own ethnic language. You are far behind compared to all the other bilingual states! Be careful, because I've heard there are some sanctions in store for you – they may, for instance, reject the allocation of the funding you requested for the space monitoring station on the Monts d'Arrée.'

The text of the recorded conversation appeared on the screen and was immediately received and translated by Schneider's fax in St. Petersburg.

Everywhere, regional languages and dialects were flourishing again, both in neo-traditional rural communities and among the urban elites. Even the Île-de-France attempted to recreate its dialects, including Parisian argot, which was now being used in various artisan guilds.

'This debate is really stupid,' Dimiri thought… With temporary 'sleep-chips' implanted in his ear, connected to a teaching robot via radio, an individual of average intelligence could learn a language of the Federation in 200 nights – just over six months. Each language cost around 230 Eurosesterces. Because of his job, Dimitri had already learned fourteen languages.

The debate in this case had taken place in French.

Other recordings followed, including the final agreement. Late at night, after some fiery arguments, the representatives of the Atlantic Arch had accepted the suggestion made by the Siberian councillor. Dimitri informed Schneider:

'Subject to confirmation on your part, I have drawn up the following plan: should we go ahead with the plan to bring food aid to North America (which would be of central geopolitical importance for us), the federal authorities could purchase large quantities of cereals, meat and milk from the plaintiff agricultural autonomous region states, in order to export them to the North American states in the grip of famine. In exchange for this, the Federation would ask these American states to accept its protectorate. Your Excellency, who has a good command of history, will realise that this would be a sort of reverse Marshall Plan.[24] This solution would help resolve the hoary controversy between the states of the Atlantic Arch and the others.'

An hour later, Schneider's laconic reply flashed on the screen:

---

24 The Marshall Plan, named after Secretary of State George Marshall, was the American reconstruction program in Europe following the Second World War.

'Report received. Brilliant solution. Suggestion accepted. Inform the Ministry of Defence.'

In contrast to the catastrophic practices of the old world, and in agreement with slogan number 65 of Vitalist Constructivism ('Like the Eagle in search of prey, politicians make decisions quickly because everything is urgent'), federal authorities reacted quickly and made clear and rapid decisions, without letting problems deteriorate or losing themselves in a labyrinth of consensus-making, consultations and commissions.

Dimitri was pleased with himself: he had done his job well. He was hoping that this time, Schneider would promote him to a higher echelon – and salary – by appointing him Plenipotentiary Legate, thus enabling him to sit on the Supreme Court for Inter-State Disputes, which solved the most difficult problems. He would add a new star – a fifth one – to his collar badges, adorned with the symbol of his corps: a silver scale on a black background, surmounted by a double-headed Eagle.

* * *

The airship made a stop at Petropavlovsk, the capital of Kamchatka. The city and its harbour sparkled with lights. In the distance, under the moonlight, stretched the Yspetsas mountain chain, from which a beam of greenish light could be seen, reaching up to the starry sky. This was the HEPL, or High-Energy Photon Line, which connected the Earth to the Cortez Moon base in the Hipparcus Crater. The line transmitted over a million megawatts of energy produced in the solar furnaces of the Moon.

The airship pitched and moored itself to the mast. Its propellers continued to turn at a reduced speed with a light humming noise. A dozen passengers took their seats. From their iron-grey uniforms, bearing the spiked wheel insignia, Dimitri recognised them to be officers from the Engineers' Legion. Among them was a tall man in a uniform adorned with the Order of the Stone Sun who greeted Dimitri with a wave. It was Engineer-General Jean-Maxime Tiernon, the man who had developed the spearhead of the armed divisions of the Federation: the Tyrannosaurus tank.

The stop in Petropavlovsk lasted no more than ten minutes. After takeoff, a steward brought passengers a light meal: smoked swordfish from the Fishing Community of the Commanders' Island, reindeer steak from the hunting tribes of Srednekolymsk and – curiously enough – some organic camembert from Normandy. The cheese had travelled quite a distance and you could tell...

There was a beeping noise. Dimitri's laptop wished to get in touch with him. He typed '18' and Vega popped up, with yet another outfit. In a tutu, she was taking a few dance steps, salacious and provocative.

'Master, His Excellency Commodore-Count Schneider has received your suggestions regarding the Brest case and approved them.'

'I know. What else, my beautiful dancing girl?'

'The High Court, in the person of Judge Kortchak, who is entrusted with negotiating with autonomous region-states that have acquired independence, is urgently asking for your opinion on the Corsican affair. He is asking me whether you suggest redeeming or invading it.'

Corsica had sought complete independence in 2059. It had been granted, following a referendum, in accordance with the Federal Constitution. But things had not gone as planned. Today it was a colony of the Sultanate of Tripoli, a deceptive and brutal regime which inflicted misery and oppression through its reign. In the meantime, a resistance movement called Corsa Libre was begging for a return of the island to the Federation.

Two months earlier, in a smart restaurant in Milan, Dimitri had discussed this problem with his friend Luigi Sutti, the Minister for Foreign Affairs of the Federation and the former President of the Parliament of the Republic of Padania.

Dimitri had made the following observation to the elegant Milanese:

'According to our informants, many Corsicans simply wish their island to be reunited with Provence. For geostrategic reasons, Corsica, which is located in the heart of the Mediterranean, cannot stay in Muslim hands. What do you think we should do?'

'Proceed through invasion and war?' Luigi Sutti had asked, sarcastically. 'We would no doubt reach our goal, but at the cost of many useless deaths. It would prove more expensive than redeeming it. The Sultan of Tripoli would be quite happy to get rid of Corsica. He needs money because of his ongoing war against the Bey of Tunis and the Islamic Republic of Egypt.'

Dimitri remembered this conversation. He had studied the case. On his keyboard he typed:

'Tell Kortchak that I'm of the following opinion: we should offer to redeem Corsica to the Sultan of Tripoli at the price of a billion

Eurosesterces. I believe he'll accept. But we should not annex Corsica to Provence. We shall make negotiations for a gradual return of the inhabitants to North Africa. We should avoid military confrontation with the Sultanate of Tripoli, who must become one of our protégés and allies in the region.'

Dimitri felt like a new Choiseul.[25]

* * *

The steward cleared the trays. He moved rather unsteadily because of the wind that was hammering the aircraft. The airship, which was now flying above the northern Pacific, appeared to be caught in a storm. Because of the greenhouse effect and the environmental catastrophes it had caused, cyclones had become increasingly common. Through the window, Dimitri could see the propellers spinning in all directions, like animals gone wild, to counteract the effect of the wind.

As was always the case in these scenarios, music was played to soothe passengers. From the loudspeakers a muted version could be heard of a popular hit by the Slovenian band Elektrock – in English, 'The Wind Blows in Gusts'. Broken by the rumble of the storm and the motors fighting to stabilise the airship, Dimitri could make out some of the words sung by Arno Magister:

The wind blows in gusts, carrying our songs
The wind blows in gusts, carrying our destinies...
Cold is our reign and the frost makes the blue steel of our swords sparkle...

The Albatross leaned to one side. One of the suitcases fell from its net. A woman started screaming. Dimitri thought of Nafissa, who was probably sleeping by now on a carriage headed to China.

The wind blows in gusts, bending the black firs.

The music suddenly stopped. All that could be heard was the wild scream of the six propellers striving to straighten the huge airship. Were they going to make it? Dimitri started praying. An advertisement about Albatrosses from the Typhoon company sprung to his mind:

---

25 Étienne François de Choiseul (1719-1785) was a French diplomat who was famous for his accomplishments. Among his achievements was the Second Treaty of Versailles, which secured Austrian support for a war against Prussia (the Seven Years War).

'Our aircraft are getting stronger and stronger.' It sounded reassuring...

Suddenly, all grew quiet. The storm had unexpectedly stopped and the airship had straightened itself out. A smiling hostess comforted the passengers by handing out glasses of *vodschkaia*.

* * *

Dimitri got back to work, but this time not with Vega. In accordance with Schneider's orders, he had to inform the Minister of Defence of the solution he had come up with at Brest concerning the purchasing of agricultural goods as a means for bringing food aid to North America.

On the screen he started reading the text of a report from the General Legation on World Information (the information services) in Berlin regarding the situation in North America. This area had never recovered from the Great Catastrophe and had broken down into many states, some of which (the central region) had completely reverted to the Middle Ages, with no traces of industrial or technological economy. Dimitri looked at a map of the region. Only four organised states still existed: the Pacific State, which was in fact a Sino-Japanese protectorate occupied by Asian military garrisons; the Old American State (OAS), the most advanced of all, which included the Great Lakes region and southern Quebec, as well as former Ontario and New England (in these two states about 9% of the population lived a technological lifestyle); the Confederation of the South, entirely agricultural and with Atlanta as its capital, which had largely sought to re-establish the Confederate way of life, stabilising its citizens' quality of life on an Eighteenth century level; and finally Dreamland (with New Orleans as its capital), a vast agricultural state in which most of the Black community had gathered following the Great Inner Exodus of 2024 – although Hispanics made up 50% of the state's population. Dreamland was beset by ongoing ethnic clashes and actually found itself as a protectorate under its Mexican neighbour, which in 2031 had quite simply annexed former New Mexico and southern California.

The rest of North America was still in the grip of chaos: communities and tribes waged ongoing mutual war amid famines and the ruins of cities and the old infrastructures. Now, the Imperial Government had received a petition from the Old American State and the Southern

Confederation. The two states were asking for emergency food relief, as climate change made agriculture extremely difficult, particularly given the return to pre-industrial methods of farming. The question was whether to send the Americans the millions of tons of flour, milk and cattle they were requesting. Of course, with the heating of the atmosphere, increased productivity in Ukraine and southern Siberia had led to a significant agricultural surplus, even now that organic farming had been adopted. But in the name of what, to help the Americans? Their petition ended with the following appeal: '...in the name of our belonging to the same civilisation.'

Dimitri remembered that some members of the Imperial Government were in favour of providing food relief for geopolitical reasons. Admiral Almagro, Baron of the Empire, Duke of Extremadura and Minister of Defence, had declared that, 'The Asian powers control the Pacific coast. Their ambition is to strengthen their presence in those areas, further to the east, and ultimately to rule North America across to the Atlantic. Would it not be in our interest to establish a protectorate over the Old American State and the Southern Confederation, in such a way as to halt this expansion? A favourable answer to the petition we have received for food aid would be a good way of extending our influence in that region. Besides, the people in these states are almost exclusively of Euro-Caucasian origin – and have been since the middle of the century. There are around 18 million of them on the whole.'

Dimitri was absorbed in the thought that the population of these two states was about five times smaller than what it had been in the Twentieth century. He immediately chose to fax a message to the cabinet of Admiral Almagro. He typed it in on his keyboard, as he didn't quite trust the microphone, fearing someone might overhear him. He thought of his career plans: the Minister of Defence would probably have appreciated the brilliant solution he had come up with, following the conflict between the states of the Atlantic Arch and the others.

Dimitri opened his text with the ritual 'Your Excellency' and then went on to describe the aim of the mission of the High Court in Brest. He concluded, 'The two parties, the states of the Atlantic Arch and the aforementioned agricultural states have reached an agreement regarding my suggestion. Federal authorities will purchase part of the produce of the plaintiff agricultural states and send it to the other side of the Atlantic as food relief. The expenses faced here will not be in the form of gratuitous funding, so to speak, but will rather serve our foreign policy plans, according to Your Excellency's views.'

Dimitri faxed the whole thing off, proud of himself, even if what he was suggesting was a small breach of the economic doctrine of 'autarchy for wide areas'.

The economic organisation of the world had, indeed, little to do now with the anarchic and catastrophic globalisation of the baleful years at the close of the Twentieth century. The Eurosiberian Federation practiced free trade within its boundaries, but outside these it was protected by extremely high customs. A bunch of bananas from the Antilles cost 90 Eurosesterces... Each great continental block lived according to its own rhythm and was economically independent. There were no longer any international flows of finance or investments.

* * *

An artificial voice announced, 'The Orthodox religious service is about to begin in the chapel on the first floor of the aircraft.'

Many people got up and took the escalator. Others chuckled. Despite the humming of the propellers and the soundproofing, Dimitri could hear snippets of songs and liturgy. 'They should thank God for having spared us from the storm,' he thought.

Dimitri was not religious, but his wife Olivia was. Following the Great Catastrophe and the expulsion of Islam from Europe, there had been a marked increase in religious practice. This hadn't benefited the Protestant Churches, which had collapsed. Catholicism had witnessed a very modest revival, hampered by the new schism and by the lack of an official Pope in Rome. In contrast, following the 2030 'Renaissance', there had been a real boom in Orthodoxy, in a bizarre form of Buddhism, and of neo-pagan cults of all sorts – from the most superstitious and wacky to the more sophisticated. The latter found inspiration in an ancient philosopher, Marcus Aurelius,[26] whose work served as a central point of reference for what might be termed 'philosophical paganism'. This current had developed a kind of syncretism between the Hellenic, Scandinavian, Germanic, Slavic and Roman traditions and was in close contact with the Hindus.

---

26 Marcus Aurelius (121-180) was a philosopher and Emperor of Rome. His *Meditations*, among other points, asserts that one must use reason to attain harmony with the cosmos.

As for Dimitri, he was both agnostic and superstitious. He believed in a sort of higher godhead indifferent towards humans, which possessed a superior intelligence and was very powerful rather than omnipotent, subdividing itself into a myriad of powers Dimitri usually referred to as the 'Devil'. Dimitri, however, was on very good terms with all religions, as required by the official ideology of Vitalist Constructivism.

* * *

There was a roar from the sky. Dimitri leaned over towards the window. Despite the darkness, he could make out a greyish, oblong and inflated object far larger than the Albatros. Some two hundred metres away, slightly above them, another airship was crossing their route.

It was a new cargo aircraft, travelling at a slow speed (200 kilometres per hour) – an eight-motor Orca. Dimitri gazed at the huge, suspended carrier, which housed the freight and cockpit. On its dark frame was a black, prancing horse on a yellow background: Ferrari. Following the disappearance of Boeing, four big companies were now vying for the world aerospace market: Ferrari, the pride of Padania; Euromotor Airbus Gesellschaft (EAG), Typhoone and Tao-Wang Air Industries. The last of these was a formidable Sino-Japanese company producing Wang-wa-sii or Flying Dragons, vacuum-filled airships that could travel at a slightly greater speed than the others. Typhoone had announced it could match them with its new 'electromagnetically suspended airships', which could reach a speed of 500 kilometres per hour and carry ten times the cargo of the old jets, while consuming ten times less energy.

The only planes now were the superlight ones of the Golden Youth. Goods were freighted via airships or ships, which ran in part on wind and hydrodynamic energy and were less polluting but just as fast. Military planes had been replaced by supersonic missile-throwing drones that could be piloted from the ground – these were known as Sharkies or 'Flying Sharks' and were produced by Typhoone – and by low-orbit satellites with powerful lasers.

The person sitting next to Dimitri, a young officer from the Engineers' Corps, addressed him:

'Do you know what they're carrying, Mr. Councillor?'

'I don't. Tell me, Lieutenant...'

'Chimeras from the bio-genetic industry in Kort. They're taking this cargo to Port Arthur.'

Chimeras were man-animal hybrids – an old dream of ancient civilisations which had become reality thanks to bio-technologies (what were now called genomics). A patent for them had been filed by two American researchers in 1998 to prevent – so the story went – these ethically shocking practices from developing any further. Chimeras ('pigmen', 'anthroporats', 'chimpanhumans') served all sorts of purposes: to produce improved sperm, as anti-rejection organ banks, as haemoglobin donors... These doped animals with human genes were filled with biotronic control chips. They were born in incubators – artificial amniotic uteruses – in the Typhoone labs in Kort, which the aircraft was flying over that very moment.

After 2050, incubators and 'supersperm' had been of great help as a means of increasing birth rates and especially improving the genetic performance of the ruling elite. Most of the population of the Federation and the world had merely reverted to the archaic demographic balance of traditional society – the age-old natural order based on high birth and death rates. As slogan no. 405 of Vitalist Constructivism stated, 'Faustianism is a form of esoterism.'

In the early Twenty-first century, following the Great Catastrophe, technological science had swept away what had been the dominant outlook for the past three centuries. Humanist and anthropocentric dogmas had collapsed. But despite this, the partisans of the old ideas enjoyed freedom of speech. On Euronet they even had a site of their own: 'The Golden Age'. The government turned a blind eye: it was good for these nostalgic old people to have a way of venting themselves.

* * *

There was a change in the speed of the propellers. 'We will be reaching Dorbisk, our final destination, in fifteen minutes,' the artificial voice said. The aircraft was gradually losing altitude. The loud speakers played a muffled version of *Douce France*,[27] a song by one Charles Trenet[28] written about a century and a half earlier.

---

27 'Sweet France'.
28 Charles Trenet (1913-2001) was a French songwriter especially popular in the 1930s, '40s and '50s.

The hostess leaned towards Dimitri. Her movements were jerky and she gave off a scent of *Ah!*, the 'ultramolecular' aphrodisiac perfume by Eros Konglomerat. Dimitri immediately realised she was a biotronic hybrid. The hostess was handing out a coloured leaflet. It was *Metamorphosis*, the official magazine of the Government, printed on glossy paper.

On the cover of the magazine was a photo of the Christopher Columbus base on Mars, which had been operational since 2062. On rocky, light-red soil, under a dirty grey-orange sky, stood some inflated or half-buried structures; next to them were men in white spacesuits seated in small vehicles with large wheels. The title read, 'On Mars we are multiplying our territory tenfold.' The article described a deal that had been signed with the Chinese Empire for the division of the Red Planet along an equatorial frontier: the north hemisphere would be left to Eurosiberia and the southern to the Chinese and Japanese. Eurosiberia's Asian rivals had thus set up a base on the south pole of the planet. Dimitri flicked through the index of the magazine. 'The Kingdom of Naples is offering rural communities ultra-resistant, low calorie maintenance work horses. The Imperial Government is signing an agreement with the Amerindian Union for the reforesting of the Amazon. The construction of the Re-Educational Penitentiary City in the Caucasus is now complete, etc.'

The Plenipotentiary Minister leafed through the magazine. The articles were replete with official slogans and techno-realist illustrations. For instance: 'Federation! Our sun never sets over our fourteen time zones,' 'The Great Homeland is not only a heritage: it's a project too,' and so on...

On a glossy interior page was an advertisement for a laser minidisc: *Our Hymns: those of our astronauts, sailors, ploughmen, lumberjacks, liberated women, etc.* Dimitri reflected that his son might have liked this – he wanted to become a musician.

Arrival

Below, Dimitri could now see his town, Dorbisk, surrounded by snow-topped hills glittering under the waning Moon, near the sparkling waters of the Bering Strait. The aircraft came to a halt and people disembarked using the lift. On the summit of the floodlit control and landing tower, the great red-and-white checkered flag of the Empire fluttered in the night, lashed by an icy wind.

Dimitri reached the entrance hall. The radio-topographic short-wave chip set in his watch informed him that Olivia was waiting for him in Hall Number Two. Thanks to the electro-biological signals from her wrist, it took Dimitri less than two minutes to find her.

'Did you have a nice day, Dimitri Leonidovich?'

'An excellent day, Olivia Fiodorovna. How're the children doing?'

'They're in bed. You'll see them tomorrow.'

She embraced him.

'I brought you a fur coat. You must be cold, coming from the warm regions of the Empire.'

Olivia covered Dimitri's shoulders with a huge wolf-fur coat.

There was a sleigh waiting for them nearby. The driver grasped the horse's reins and the snow started crunching under the sleigh's runners. Their house was only ten minutes away from the airport.

In the main room of the house, a large peat fire gave off pleasant, scented and sweet heat.

As Dimitri sat in front of the fireplace, Natcha, his young maidservant, served him a platter of raw fish marinated in a sour wild-nettle sauce – a traditional Siberian dish.

Olivia watched her husband eat with her large blue eyes and a questioning, almost anxious air.

'Did you accomplish your mission?'

'Yes.'

'Are we going to spend fifteen days holiday together, then?'

'Yes.'

'Did you see, Dimitri? The sun is rising.'

Beyond the wooden frame of the window, light shone from the east. Far off, the snowy peaks of Alaska were visible, enveloped in the morning mist. In the violet sky, a musical roar and a fast-moving streak of smoke revealed the presence of a Sharkie 27 – the aeronautical pride of the Typhoone company. At Mach 7, 25,000 metres above the ground, it crossed the icy sky. The stratospheric patrols of these flying sharks were securing the frontiers of the Empire.

Dimitri unpacked and gave Olivia the jewel he had brought for her from Brittany for their ten-year anniversary.

'Come, let's go to sleep.'

A DAY IN THE LIFE OF DIMITRI LEONIDOVICH OBLOMOV

Facing the bed was a painting by the Twentieth century French artist Olivier Carré. It was a small green-and-grey oil canvas entitled *Fin*,[29] with a steel frame that the artist had made himself. The painting depicted a monster, 'Le Grand Albert'.[30] His eyes appeared red and threatening, although there was no red in the picture. It was dated 1982.

Half asleep, Dimitri could hear his children laughing from the room upstairs. The white radiance of the Siberian sun always woke them up early.

The last image Dimitri Leonidovich Oblomov saw before his eyes before falling asleep was the huge red-and-white checkered flag – the living symbol of the Great Homeland. Red: like the blood shed and the blood it protected and served; white: like the radiance of the rising sun, like pure strength and loyalty.

* * *

All the scientific information provided in this story is accurate and not merely the product of the author's literary imagination. For the inventions described, patents have been filed in the late Twentieth century. They were only developed later, however, in the Archeofuturist age, from a very different perspective...

---

29 French: 'end'.
30 'Albert the Great' was Saint Albertus Magnus (1193?-1280), a Dominican bishop who attempted to reconcile science and religion. He was also noted for being the first Medieval thinker to merge Aristotle with the Catholic tradition.

Other books published by Arktos:

Beyond Human Rights
by Alain de Benoist

The Problem of Democracy
by Alain de Benoist

Revolution from Above
by Kerry Bolton

Metaphysics of War
by Julius Evola

*The Path of Cinnabar:
An Intellectual Autobiography*
by Julius Evola

Why We Fight
by Guillaume Faye

The WASP Question
by Andrew Fraser

The Saga of the Aryan Race
by Porus Homi Havewala

The Owls of Afrasiab
by Lars Holger Holm

De Naturae Natura
by Alexander Jacob

Can Life Prevail?
by Pentti Linkola

A Handbook of Traditional Living
by Raido

The Jedi in the Lotus: Star Wars and the Hindu Tradition
by Steven J. Rosen

It Cannot Be Stormed
by Ernst von Salomon

Tradition & Revolution
by Troy Southgate

Against Democracy and Equality: The European New Right
by Tomislav Sunic

The Initiate: Journal of Traditional Studies
by David J. Wingfield (ed.)

CPSIA information can be obtained
at www.ICGtesting.com
Printed in the USA
JSHW012205110123
36143JS00006B/200/J